Single-Inductor Multiple-Output Converters

Single-Inductor Multiple-Output Converters
Topologies, Implementation, and Applications

Albert Ting Leung Lee
Weijian Jin
Siew-Chong Tan
Ron Shu Yuen Hui

CRC Press
Taylor & Francis Group
Boca Raton London New York

CRC Press is an imprint of the
Taylor & Francis Group, an **informa** business

First edition published 2022
by CRC Press
6000 Broken Sound Parkway NW, Suite 300, Boca Raton, FL 33487-2742

and by CRC Press
2 Park Square, Milton Park, Abingdon, Oxon, OX14 4RN

CRC Press is an imprint of Taylor & Francis Group, LLC

Library of Congress Cataloging-in-Publication Data
Names: Lee, Albert Ting Leung, 1972- author. | Jin, Weijian, author. |
Tan, Siew-Chong, author. | Hui, Ron, 1961- author.
Title: Single-inductor multiple-output converters : topologies,
implementation, and applications / Albert Ting Leung Lee, Weijian Jin,
Siew-Chong Tan, Ron Shu Yuen Hui.
Description: First edition. | Boca Raton : CRC Press, 2022. |
Includes bibliographical references and index. |
Summary: "The book provides a comprehensive overview of Single-Inductor Multiple-Output Converters from both theoretical and practical perspectives. Based on the authors' in-depth research, the volume covers not only conventional SIMO DC-DC converters but also the new generations of SIMO such as SIMO AC-DC converters, SIMO DC-AC converters, and the latest SIMO hybrid con-verters"—Provided by publisher.
Identifiers: LCCN 2021027405 (print) | LCCN 2021027406 (ebook) |
ISBN 9781032145358 (hbk) | ISBN 9781032145389 (pbk) |
ISBN 9781003239833 (ebk)
Subjects: LCSH: Single-inductor multiple-output converters.
Classification: LCC TK7872.C8 L43 2022 (print) | LCC TK7872.C8 (ebook) |
DDC 621.3815/322—dc23
LC record available at https://lccn.loc.gov/2021027405
LC ebook record available at https://lccn.loc.gov/2021027406

ISBN: 978-1-032-14535-8 (hbk)
ISBN: 978-1-032-14538-9 (pbk)
ISBN: 978-1-003-23983-3 (ebk)

DOI: 10.1201/9781003239833

Typeset in Minion
by codeMantra

Access the Support Material: https://www.routledge.com/9781032145358

Contents

CHAPTER 1

Introduction to Single-Inductor Multiple-Output Converters

1.1 INTRODUCTION TO SIMO TOPOLOGY

To our best knowledge, the advent of single-inductor multiple-output (SIMO) DC-DC converters can be dated as far back as 1997, the year when a U.S. patent application of a SIMO boost regulator was filed, and subsequently, a US patent was published on June 13, 2000 [1]. Soon afterwards, an increasing number of conference papers and journal articles related to SIMO DC-DC converters were published [2–18]. Many electronic devices or systems require multiple regulated supply voltages for different function modules. Dynamic voltage scaling (DVS) is a common technique to reduce the average power consumption in embedded systems. DVS is typically used in battery-operated portable electronic devices such as mobile phones, tablets, wearables, and sensors where power savings and battery life are paramount. It is also used in large systems with multiple microprocessors or digital signal processors where power saving is needed for thermal management.

DOI: 10.1201/9781003239833-1

Conventionally, to provide multiple supply voltages, a straightforward approach is to use multiple independent switching converters. In general, N single-output switching converters are employed to generate N regulated output voltages. This requires a total of N inductors and $2N$ power devices (power MOSFETs and diodes). Figure 1.1 shows the system architecture of a conventional implementation using multiple DC-DC switching converters. With the advancement of semiconductor technologies, power MOSFETs, gate drivers, and control circuits are usually integrated. Unfortunately, off-chip inductors are still the bulkiest and perhaps, one of the most expensive components in switch mode power supply (SMPS), which pose a huge challenge in the miniaturization of DC-DC converters. On the other hand, to provide galvanic isolation between the input and multiple outputs of an isolated multiple-output flyback converter with a single controller, a transformer with one primary winding and multiple secondary windings can be used. In most cases, the voltage of the master output, typically the one with the heaviest load, is tightly regulated via closed-loop control while the other outputs are loosely controlled (or "quasi-regulated") through coupling of the secondary windings. It is susceptible to cross-regulation as a change in the master output will also affect the other outputs. Yet both implementations inevitably require the use of bulky magnetics, which are difficult to integrate. Driven by a strong

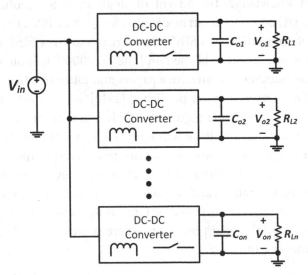

FIGURE 1.1 System architecture of a conventional implementation using multiple DC-DC switching converters.

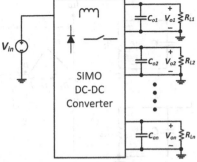

FIGURE 1.2   System architecture of a SIMO DC-DC switching converter.

desire to achieve small size, light weight, and low cost for multiple-output converters, considerable research efforts have been devoted to exploring new circuit topologies using only one inductor to supply multiple outputs. Ultimately, this has led to the invention of a new circuit topology called SIMO. The use of a single inductor to drive multiple unidentical loads has quickly emerged as a highly scalable and promising solution for space-constrained and cost-conscious applications. Figure 1.2 shows the system architecture of a SIMO DC-DC converter.

As an example, let us consider two conventional DC-DC buck switching converters, namely, Converter A and Converter B, with the same switching frequency f_s. Both converters operate in discontinuous conduction mode (DCM). The on-time period of the power switch of the two converters is denoted by $D_{1a}T$ and $D_{1b}T$, respectively, where T_s is the switching period ($T_s = 1/f_s$). Suppose the two converters work in two complementary phases ϕ_a and ϕ_b in such a way that $D_{1a}T + D_{2a}T < 0.5$ and $D_{1b}T + D_{2b}T < 0.5$. Figure 1.3 depicts the circuit diagrams and the corresponding waveforms of the inductor currents of the two buck converters in DCM.

For Converter A, the inductor current ramps up during the first subinterval ($D_{1a}T$) as the inductor is charged. It ramps down during the second subinterval ($D_{2a}T$) as the inductor is subsequently discharged. Finally, it returns to zero during the third subinterval ($D_{3a}T$), where $D_{3a}T = 1 - D_{1a}T - D_{1b}T$. DCM is characterized by the inductor current being zero in the third subinterval (the so-called idle phase). Likewise, Converter B also has a similar switching sequence. Since the power switches S_a and S_b are enabled in ϕ_a and ϕ_b, respectively, the energy is diverted to the two outputs V_{oa} and V_{ob} in separate time intervals. As a

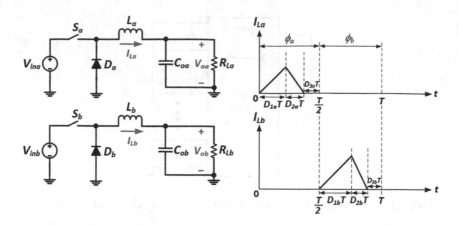

FIGURE 1.3   Circuit diagrams and inductor current waveforms of the two buck switching converters.

consequence, the operations are fully decoupled in the time domain and thus independent of each other. Hence, the two buck converters can essentially be combined into a single buck converter whose circuit diagram and inductor current waveform are depicted in Figure 1.4.

Figure 1.4 shows that two complementary output switches S_a and S_b are used to select which output to connect to the main inductor L. It is crucial to note that S_a and S_b *cannot* be simultaneously switched on under all circumstances. S_a is switched on only during the first phase ϕ_a, while S_b is switched on only during the second phase ϕ_b. In other words, the energy stored in the inductor is transferred to the two outputs successively in a round-robin time-multiplexing (TM) manner. The so-called TM switching scheme enables the sharing of a single inductor between the two outputs without any cross-channel interference, which is also referred to as

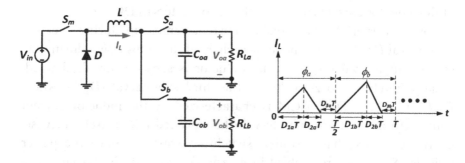

FIGURE 1.4   Circuit diagram and inductor current waveform of SIDO buck converter operating in DCM.

cross-regulation. In this way, a single-inductor dual-output (SIDO) converter is obtained. For ease of modeling and analysis, the SIDO converter operating in DCM can be functionally decomposed into two individual subconverters. The feedback controller of each subconverter can then be designed separately.

Naturally, we can extend this TM switching scheme to an arbitrary number of outputs. Each output will be allocated a uniform time slot of T/N for the charging and discharging of the inductor, where N is the number of outputs. Simply put, the inductor is being charged (or discharged) N times within a full period T. Such switching scheme is referred to as the multiple-energizing method. Hence, this gives rise to a whole new family of SIMO converters.

Nonetheless, a major drawback of operating the SIMO converter in DCM is that the output current is relatively small. This implies that the output power that can be delivered by the SIMO converter is rather limited. In DCM, the only way to achieve a higher output current is to increase the peak value of the inductor current at a fixed switching period. But this inevitably leads to an increase in the current stress of the inductor and power MOSFET(s). Ultimately, the maximum peak current is reached when the SIMO converter operates at the boundary conduction mode (BCM) or critical conduction mode, which is a boundary condition between continuous conduction mode (CCM) and DCM. To address the issues of power limitation and increased current stress, a new operating mode named pseudo-continuous conduction mode (PCCM) is introduced primarily for SIMO switching converters [5]. In PCCM, the floor of the inductor current is raised by a DC level of I_{dc}, which effectively increases the average value of the inductor current. This is made possible with the addition of a freewheel switch across the inductor. Figure 1.5 shows a SIDO converter and the corresponding inductor current waveform in PCCM. To meet the high current demand of heavy loads, I_{dc} can be increased to allow more power to be delivered to the outputs. Compared to DCM, the inductor current ripple ΔI_L can be much reduced as a larger inductor can be used in PCCM. Thus, the power constraints incurred in DCM can be safely eliminated, and the converter can achieve relatively small current and voltage ripple. In short, by operating the SIMO converter in PCCM, the output power can be increased while maintaining zero cross-regulation. Chapter 2 focuses on SIMO DC-DC converters and their operating principle.

The predecessors of SIMO converters are designed specifically for DC-DC power conversion. From a single DC power source (e.g., a battery cell), they can produce multiple DC output voltages by employing only a single inductor in the power stage. A number of on-chip SIMO DC-DC boost converters with only one off-chip inductor were reported in the literature [3–7]. Compared with the traditional multiple-output converter topologies, these SIMO converters require fewer inductors, power devices, and control loops, which carry the advantages of small form factor, reduced component count, low build-of-material cost, better scalability to multiple outputs, and high efficiency. The SIMO DC-DC converters can be used in a number of key applications, including, but not limited to, voltage regulator modules, system-on-chips (SoCs), and embedded power supplies for a myriad of mobile devices. In addition, SIMO DC-DC converters can also act as SIMO LED drivers for supplying multiple independent LED strings. This helps extend the scope of applications of SIMO, which include LED backlighting for LCD displays as well as ambient lighting and intelligent color-mixing applications via red-green-blue (RGB) LED lighting. In high-power lighting applications such as streetlight, commercial or residential lighting, and large-scale LCD panels, offline AC-DC LED drivers are frequently used to drive multiple parallel LED strings. These are high-power LED drivers that can operate directly from the AC mains, e.g., 110 VAC/60 Hz or 220 VAC/50 Hz. Conventional topologies for driving AC-DC multistring LED systems have been proposed, which can generally be classified into two types as shown in Figure 1.6a and b. Figure 1.6a shows an AC-DC stage which generates a single common output bus that is shared by all the LED strings connected in parallel [19–21]. In comparison, Figure 1.6b shows an AC-DC stage that generates a separate output voltage for each LED string [22–24].

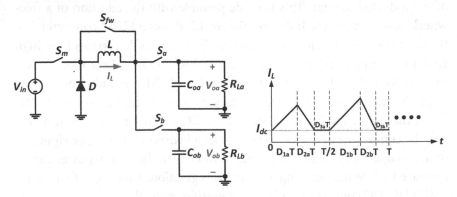

FIGURE 1.5 Circuit diagram and inductor current waveform of SIDO converter operating in PCCM.

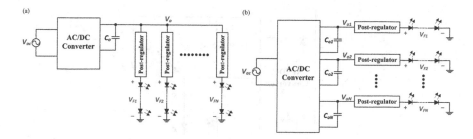

FIGURE 1.6 Conventional topologies of offline AC-DC LED drivers with (a) a common output bus for all LED strings and (b) a separate output bus for each individual LED string.

To enable independent current regulation of each individual string, the AC-DC preregulation stage is cascaded with a parallel combination of postregulators. Each postregulator is responsible for regulating the current of the string to which it is connected. In general, there are two types of postregulators, namely, linear type and DC-DC converter type. Even though linear postregulators have a simple circuit structure, they may suffer from significant power loss if not properly designed. DC-DC postregulators are ideally lossless, but each postregulator introduces additional power switches, inductors, and other passive components to the overall system. Unfortunately, as the number of LED strings increases, the size and cost of the LED system will also increase. These conventional two-stage topologies also require two sets of controllers, one for the AC-DC stage and another for the postregulators, which increases the controller complexity. Besides, they require the use of a large DC-link capacitor, typically a high-voltage electrolytic capacitor (E-cap), in order to mitigate the double-line-frequency ripple. If the voltage of the intermediate DC-link is high, it becomes quite challenging to choose a larger and yet inexpensive capacitor that has a long lifetime. The use of a short-lifetime electrolytic capacitor would undermine the reliability of LED drivers [25].

In view of the aforementioned issues, an offline single-stage AC-DC SIMO LED driver is introduced for medium- to high-power multistring LED applications [26]. It is capable of achieving both functions of AC-DC rectification with a high power factor and precise and independent current regulation of each individual LED string. Figure 1.7 shows the system architecture of a single-stage AC-DC SIMO LED driver. Chapter 3 provides a detailed description of the AC-DC SIMO LED driver. Indeed, this is a major breakthrough in SIMO technology because it shows that the

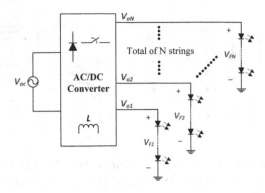

FIGURE 1.7   System architecture of a single-stage AC-DC SIMO LED driver.

concept of SIMO can be extended beyond DC-DC to AC-DC converters. Subsequently, this raises yet another question of whether the SIMO architecture can also be used to realize DC-AC power conversion. This book will provide answers to this question.

The conventional DC-AC converter (or inverter) topologies for generating multiple AC outputs require either multiple independent power inverters [27–29] or a multistage power conversion architecture with sophisticated control methods [30–32]. Figure 1.8 shows the system architecture of a conventional bridge-type DC-AC power supply with multiple independent AC outputs. Basically, the power stage consists of separate full-bridge phase-shift inverters and a parallel LCL resonant network.

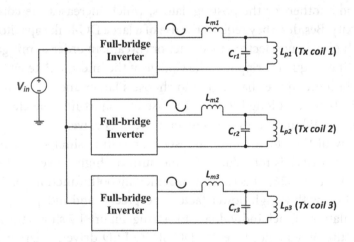

FIGURE 1.8   System architecture of a conventional DC-AC power supply with multiple AC outputs.

The full-bridge inverters are supplied by a single DC power source such as an external battery, a USB port, or the output of a DC-DC converter. The LCL resonant tank comprises the matching inductor, the resonant capacitor, and an inductive load, e.g., a transmitting coil for wireless power transfer. Obviously, a major drawback of this simplistic topology is that the number of inverters increases proportionally with the number of AC outputs. This unavoidably leads to a bigger form factor, reduced power density, lower efficiency, and higher cost as the number of AC output increases. Another major issue is the lack of coordination or synchronization among the AC outputs due to the independent operations of the inverters.

On the other hand, a three-stage power conversion architecture for driving multiple AC outputs is shown in Figure 1.9. The first stage comprises a DC-AC converter, which is typically implemented as a full-bridge inverter. The center stage is a power demultiplexer, which selects one of the output resonant circuits to be connected to the preceding inverter stage. The final stage is a parallel connection of series resonant circuits. Although this approach requires only one inverter, the implementation of the power demultiplexer is more sophisticated, which requires a multitude of discrete relays such as electromechanical relays, solid-state relays, or FET switches. It is conceivable that as the number of AC output increases, more discrete relays are needed, thereby resulting in a larger size, increased power loss, and higher cost. Moreover, this three-stage architecture requires two separate sets of controllers, one for the full-bridge inverter and the other for the

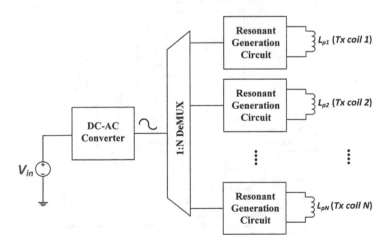

FIGURE 1.9   System architecture of a three-stage DC-AC power conversion topology.

power demultiplexer. A greater number of AC loads inevitably complicates the controller design. Therefore, there is a stronger need to come up with a compact, low-cost, and efficient DC-AC power supply for practical applications demanding a flexible number of unidentical AC loads.

In light of the abovementioned issues, a whole new class of SIMO-based inverter topologies has recently emerged [33–36]. It enables the use of a single inductor to simultaneously deliver energy to multiple AC loads. Fundamentally, this topology is formulated by integrating the function of a conventional DC-DC SIMO topology and that of a DC-AC stage (i.e., the resonant tank) into a single stage. Figure 1.10 shows the system architecture of the SIMO inverter with closed-loop control. This SIMO inverter is superior to the conventional multioutput inverters because it requires fewer power switches, gate drivers, and passive components while attaining robust DC-AC power conversion, enhanced scalability to multiple AC outputs, and high power efficiency. No isolated transformer is required, which significantly reduces the overall size and weight, and thus increases the power density. More importantly, this inverter achieves one-stage DC-AC power conversion, and it employs only one controller to simultaneously regulate all AC output voltages. It results in a simple and scalable control scheme. Hence, the SIMO inverter is compact, efficient, and cost-effective. In practice, it can serve as a multicoil wireless power transmitter (or wireless charger) for quick, concurrent, and convenient wireless charging of multiple portable electronic

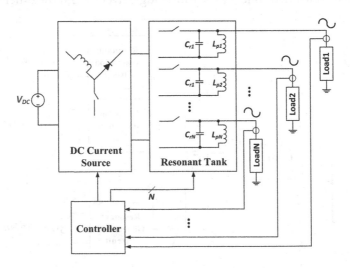

FIGURE 1.10 System architecture of SIMO inverter with closed-loop control.

devices. Chapters 4–6 present three distinctive topologies of SIMO inverters alongside the theoretical underpinning and design methodology.

Last but not least, Chapter 7 introduces the latest addition to the family of SIMO, namely, SIMO DC-AC/DC hybrid converters. In brief, this hybrid converter can simultaneously supply both AC and DC loads from a single DC power source. This hybrid topology is capable of achieving any combination of AC and DC outputs. Besides, standalone AC (or DC) operation is allowed. It can act as a practical high-capacity hybrid charger or power bank that enables quick and simultaneous wireless and wired power transfer.

1.2 SWITCHING SCHEME

Since there is only one main inductor in the power stage of a SIMO topology, which is shared by all outputs, proper switching sequence of the power switches is needed to ensure that the main inductor has sufficient energy to supply all the output loads. Therefore, the main objective of a switching scheme is to provide a continuous and stable power delivery across the individual outputs and meet their power demands. In general, the switching scheme of SIMO converters (or inverters) can largely be divided into two major types: the multiple-energizing method and single-energizing method. The term "energizing" refers to the charging of the main inductor. Likewise, "de-energizing" refers to the discharging of the main inductor.

1.2.1 Multiple-Energizing Method

As the name suggests, the multiple-energizing method means that the main inductor is being charged *multiple* times per switching cycle. This switching method was originally proposed for the early versions of SIMO DC-DC converters, and it still remains the most popular switching scheme of SIMO-based power supplies today. Building upon the concept of TM, the energizing and de-energizing of the main inductor are performed on a "per-output" basis in every switching cycle. Figure 1.11 provides a graphical illustration of the multiple-energizing switching method. It shows the theoretical waveforms of the inductor current of a SIMO converter operating in three different modes of operation, namely, CCM, DCM, and PCCM. Regardless of the mode of operation, a full operating period T is partitioned into N equal switching cycles, and only one output is enabled in each switching cycle.

A traditional single-input single-output DC-DC converter can operate in either CCM or DCM. We may be inclined to think that a multiple-output

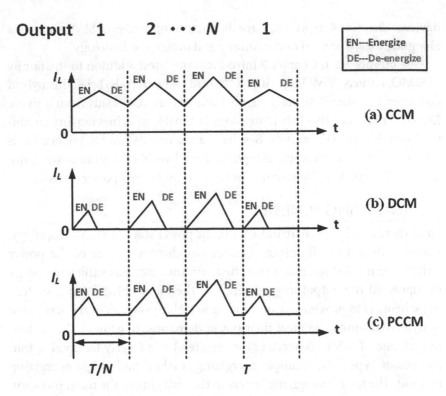

FIGURE 1.11 Theoretical waveforms of the inductor current of a SIMO converter operating in (a) CCM, (b) DCM, and (c) PCCM under the multiple-energizing switching method.

converter like SIMO can also operate in either CCM or DCM. In CCM, the utilization of the switching cycle is maximized, which implies that it can deliver more power. In DCM, the third subinterval (idle phase) is purposefully left unused so that the outputs are fully decoupled in the time domain, which helps mitigate the cross-regulation issue. Yet, this idle time leads to an underutilization of the switching cycle that limits the charging time of the main inductor. Less energy is being stored in the main inductor, which in turn results in smaller output power. To resolve the seemingly conflicting requirements of higher output power and minimum cross-regulation, a new operating mode named PCCM is subsequently introduced [5]. In PCCM, the inductor current has a positive DC offset, which effectively increases the average value of the inductor current. Hence, a higher output power can be achieved with zero cross-regulation. As shown in Figure 1.11, regardless of the modes of operation, the main inductor is energized and de-energized *once* for each individual output within a switching cycle. Simply put, the SIMO converter will serve only one output per cycle. If there are a total of N outputs, the main inductor will be charged and discharged N times in a row. In other words, it will take the SIMO converter N switching cycles to finish energizing *all* outputs. The switching frequency of the main switch is therefore N times that of each output switch.

Perhaps the most salient feature of the multiple-energizing switching method is that the initial state of the main inductor in DCM (or PCCM) is the same across all outputs. This is because the inductor current always returns to zero in DCM (or a constant DC offset in PCCM) at the beginning of every switching cycle as illustrated in Figure 1.11. Imagine what might happen if, as a particular output load demands more power, the peak of the inductor current suddenly increases. In the case of DCM (or PCCM), a higher peak current will undoubtedly shorten the third subinterval (idle phase) at a fixed switching period. But as long as there is still sufficient time for the inductor current to return to zero value in DCM (or a constant value in PCCM), it will *not* affect the initial state of the next cycle. Hence, cross-regulation can be avoided. In the event that the SIMO converter reaches a boundary condition between DCM and CCM, commonly referred to as BCM or critical conduction mode, a further increase in the peak current would result in cross-regulation. To eliminate cross-regulation in this case, one option is to increase the full period T, which in turn increases the switching cycle. This will give adequate time for the inductor current to return to zero, thus reinstating DCM operation. However, since the frequency of energy supply per output decreases, the output voltage ripple will become larger. Consequently, a larger output capacitor is required so as to maintain a reasonably small output ripple.

1.2.2 Single-Energizing Method

Instead of charging and discharging the main inductor multiple times within a full period T, it is also possible to energize (or charge) the main inductor *only once* and then distribute the energy stored in the inductor to each output successively. Such switching scheme is referred to as the single-energizing method. The switching sequence of a SIMO converter using the single-energizing method is graphically illustrated in Figure 1.12. It shows the theoretical waveforms of the inductor current of a SIMO converter operating in CCM, DCM, and PCCM, respectively. The main consideration of this particular switching method is to be able to accurately predict

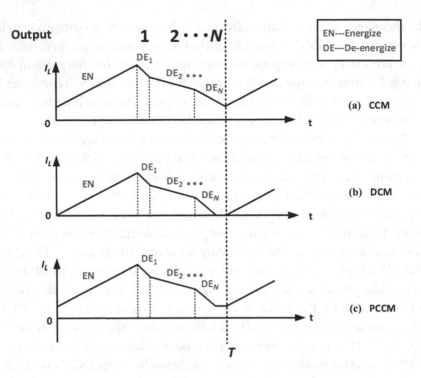

FIGURE 1.12 Theoretical waveforms of the inductor current of a SIMO converter operating in (a) CCM, (b) DCM, and (c) PCCM under the single-energizing switching method.

the total power demand of all outputs beforehand. This sets the charging time of the main inductor or the total amount of energy that needs to be stored in the inductor. The discharging time of each output is then determined by its own power consumption. Thus, the switching frequency of the main switch is the same as that of the output switch.

It is worth mentioning that there is a major limitation of this single-energizing switching method. Unlike its multiple-energizing counterpart, there is no idle time between adjacent discharging cycles in the single-energizing method to decouple the outputs in the time domain. Similar to CCM, the inductor current flows continuously from one output to the other, which renders the SIMO converter vulnerable to cross-regulation. Imagine what might occur if there is a sudden load increase (or decrease) in the first output while the loads of the other outputs remain constant. Since the total energy of the inductor is unchanged in a particular cycle, more energy being drawn by the first output means that the energy left in the inductor, already depleted by the other outputs, may be insufficient to sustain the load of the last (N^{th}) output. This cumulative effect can also be understood in the time domain. Let us refer to Figure 1.12 again. The peak value of the inductor current at a particular cycle is assumed to be fixed and so does the switching period T. Suppose there is a load *increase* at the first output, which translates to a *longer* discharging time interval for the first output. By using simple geometry, a *longer* discharging interval for the first output will ultimately result in a *shorter* discharging interval for the last output, assuming that there is still sufficient time left to complete the discharging cycles of all the intermediate outputs, i.e., between the second output and the $(N-1)^{th}$ output. But if the load increase of the first output becomes so large that there is insufficient energy remaining in the inductor to sustain the last output (or inadequate discharging time for the last output before the inductor current drops to zero), the output voltage of the last output will decrease. Thus, a load increase at the first output could potentially perturb the load at the last output. It is therefore conceivable that the single-energizing method is prone to cross-regulation. Perhaps, one plausible solution is to increase the peak value of the inductor current accordingly in order to accommodate the load increase of the first output. But this requires a more sophisticated control scheme that can respond precisely and quickly to the load transient. Besides, the use of a higher peak current value tends to increase the current stress of the main inductor and power MOSFETs, which may not be practical. The fact that the fixed switching period limits the maximum peak value of the inductor current sets an upper bound to the output power that can be delivered to the outputs. Nonetheless, despite its drawbacks, the single-energizing method carries one significant advantage. It allows the SIMO converter to operate in burst mode. Due to the absence of idle periods between adjacent outputs, the single-energizing method can achieve a better utilization of the switching cycle, which enables the use of a higher switching frequency. Hence, by operating at a higher switching frequency (i.e., the switching cycle becomes smaller), each output of the SIMO converter receives less energy from the main inductor per cycle because of shorter discharging interval but the frequency of energy supply actually increases. A smaller output capacitor in each output can therefore be used and yet a small output voltage ripple can still be attained.

More recently, a hybrid switching scheme is introduced, which combines the multiple-energizing method with the single-energizing method [37]. In a single-inductor three-output (SITO) inverter with three distinct output frequencies, the multiple-energizing switching method would hypothetically schedule two adjacent outputs in the fifth cycle, which allows them to receive uninterruptible and uniform energy supply from the power stage of the inverter. However, doing so will create a time conflict as both outputs would like to operate within the same switching cycle, hence violating the principle of the multiple-energizing method. To resolve this time conflict, the single-energizing method is employed specifically in the fifth cycle. Basically, it allows both outputs to be enabled sequentially only during the fifth cycle. The multiple-energizing method continues to be used in the rest of the switching cycles. One example of the hybrid switching scheme is called the modified Type III switching sequence, which will be elucidated in Chapter 6.

1.3 GENERAL APPLICATIONS

The versatility, scalability, and flexibility of the SIMO topology enable it to be used in a broad range of practical applications. For example, SIMO DC-DC converters are used as the power management unit (PMU) in a myriad of battery-powered applications including SoCs and portable electronic devices. Basically, SoC integrates the whole system onto a single integrated circuit (IC) to achieve high integrations. At the chip level, PMU is an essential part of the system that provides multiple regulated voltage supplies to analog and digital circuits within the SoC. Since the SIMO topology requires very few external components, it is particularly suitable for realizing PMU to attain high compactness, increased power density, low cost, and high efficiency. At the board (system) level, the power management IC (PMIC) plays an important role in various kinds of battery-operated mobile devices such as mobile phones, tablets, laptops, hearables, wearables, sensors, and other Internet-of-Things (IoT) gadgets. The microprocessors or microcontroller unit (MCU) and other peripheral ICs in these devices need to be driven by multiple tightly regulated low-voltage DC supply voltages to ensure robust and optimal performance. Hence, a low-cost and highly efficient PMIC in a single IC package with small footprint can be achievable by the SIMO architecture. Figure 1.13 shows a typical power supply system of a wearable IoT device using a SITO PMIC. The SITO DC-DC converter transforms the 3.8 V DC input voltage into three independently regulated DC output voltages, i.e., 1.2, 2.05, and 3.3 V, respectively.

Besides embedded power supplies, the SIMO topology also finds applications in practical LED lighting systems such as LED backlighting for

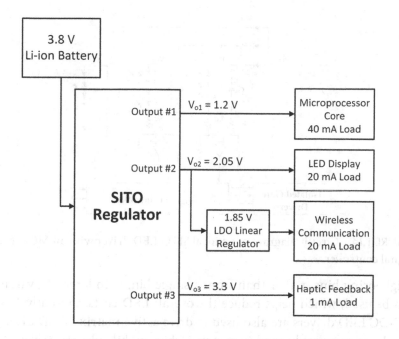

FIGURE 1.13   Power supply system of a wearable IoT device using SITO PMIC.

LCD displays, torch or flashlight, and ambient lighting. For instance, a SITO DC-DC converter acts as a RGB LED driver, which transforms a single DC input voltage into three independently regulated DC currents (I_R, I_G, I_B), where I_R is the current across the red LED string, I_G is the current across the green LED string, and I_B is the current across the blue LED string. In reality, the DC input voltage can be supplied by a high-capacity battery pack, portable power bank, or even the intermediate DC-link voltage in an AC-DC LED system. Figure 1.14 shows a circuit diagram of an ideal SITO LED driver with a digital controller like a low-cost MCU.

Rather than using a common bus voltage for all LED strings in conventional LED drivers, the SIMO LED driver enables the optimization of the individual output voltage to compensate for the variability of the LED forward voltage V_F in each of the different colored LED strings. This allows the output voltage to closely match the total V_F per LED string, which leads to a much smaller voltage headroom and hence, reduces the power loss. Moreover, because each output voltage can be independently controlled in the SIMO LED driver, loosely binned LEDs with larger variation of V_F can therefore be used. In other words, same-colored LEDs from

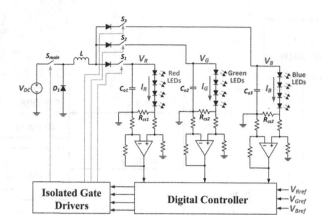

FIGURE 1.14 Circuit diagram of an ideal SITO LED driver with an MCU-based digital controller.

neighboring bins, rather than from a single bin, with larger V_F variance can be used, which helps reduce the overall LED costs. Presently, SIMO DC-DC LED drivers are also used to drive active-matrix OLED displays, which are commonly used in smartwatches, mobile phones, tablets, laptops, monitors, and TVs. The SIMO-based solution enables the LED driver to be compact, light-weighted, and highly efficient to achieve miniaturization of mobile devices and maximization of battery operating lifetime.

Since most of the existing commercial and residential lightings are supplied directly by the AC mains, an offline AC-powered SIMO multichannel LED driver is introduced. This is a single-stage AC-DC LED driver with a single AC input and multiple DC outputs whose schematic diagram is shown in Figure 1.15. It enables the direct conversion of a universal AC supply into multiple independently controlled DC output currents for driving multiple LED strings. This AC-DC SIMO LED driver can be used across a wide spectrum of high-power lighting applications, including, but not limited to, theaters, museums, studio lighting, and architectural lighting. Owing to its compact size, the AC-DC LED driver based on the SIMO topology can easily fit into the existing luminaires casing (e.g., MR16 or E27 light bulb, T5/T8 light tube) for various LED retrofit applications.

Nowadays, wireless power technologies have been gaining traction in a wide range of practical applications, which include the charging of low-power electronic devices such as smartphones, tablets, wearables, and biomedical implants to high-power charging of electric vehicles. The multicoil wireless power transfer (MCWPT) system emerges as an increasingly

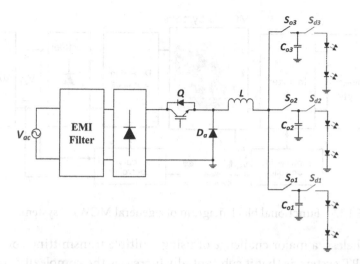

FIGURE 1.15 Schematic diagram of an AC-powered SIMO multichannel LED driver.

attractive solution due to its flexibility, scalability, and cost-effectiveness of charging multiple devices concurrently. Figure 1.16 shows an inside view of a Qi-compliant MCWPT system. Figure 1.17 shows a functional block diagram of a general MCWPT system.

The use of multiple coils increases the likelihood that at least one of the transmitting coils will be aligned closely with the receiving coil of the device being charged. It significantly improves the freedom of positioning of the receiving device on the wireless charging pad. Besides, a multicoil wireless transmitter makes it possible to charge multiple electronic devices simultaneously. Imagine only one charging pad is needed for concurrently charging various types of Qi-enabled devices. Another advantage of a multicoil system is that it facilitates adaptability and scalability of the transmitter to better cope with multiple unidentical devices that are rated at different power levels and charging speed. It allows flexibility and scalability of the charging area without sacrificing power transfer efficiency.

FIGURE 1.16 An inside view of a Qi-compliant MCWPT system.

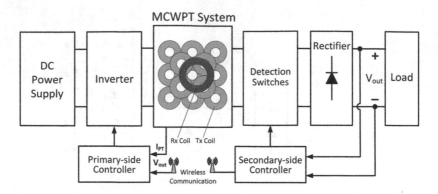

FIGURE 1.17   Functional block diagram of a general MCWPT system.

Nonetheless, a major challenge of using multiple transmitting coils in an MCWPT system is that it substantially increases the complexity and cost of the transmitting design. A key component of the wireless power transfer system is the power inverter which drives the transmitting coil. The inverter converts a DC input supply into a high-frequency AC current to energize the transmitting coil, which generates an oscillating magnetic field. To realize an MCWPT system, a conventional approach is to use multiple inverters, one for each transmitting coil. However, a major drawback of such multiple-inverter topology is that the number of inverters increases proportionally with the number of transmitting coils. This significantly increases the overall size, cost, and design complexity of the system. The large form factor of the inverters can easily limit MCWPT from being used in space-constrained applications. Its high cost also prevents the wide adoption of wireless power in general applications. Primarily driven by the fast-growing market demand and more stringent design constraints, there is a strong motivation in R&D works to develop a compact, efficient, and low-cost inverter for future generations of MCWPT systems. Nevertheless, the design of such inverters poses two new challenges.

- Existing inverter topologies for simultaneously charging multiple devices require an excessive number of power switches and passive components, which are bulky, costly, and inefficient. These inverters are unacceptable for many space- and cost-stringent applications. In addition, they only support charging of a limited number of devices at a time because of relatively low power ratings. This is insufficient to satisfy the ever-increasing power demands from many newer types of consumer electronic devices. To attain truly cost-effective and high-power-density inverters for medium-to-high power wireless power transfer, we need to explore new topologies that use fewest components while achieving a wide output power range.

- Cross-coupling among transmitting coils in the MCWPT system has always been a major issue. It becomes more pronounced when the transmitting coils are placed in close proximity to one another. Such problem is observed when the transmitting coils are oscillating at the same frequency. Hence, it is essential to operate the transmitting coils at different frequencies in order to mitigate the undesirable cross-interference between coils.

The aforementioned challenges lead to the proposal of novel SIMO inverter topologies for future MCWPT systems, which are capable of charging multiple unidentical devices concurrently with fast and stable dynamic performances at higher power ratings. Qi is the world's de facto wireless charging standard for providing a low power (5–15 W) to small personal electronics [38]. At the time of writing, the Wireless Power Consortium is also working on a medium power standard that delivers wireless power to products operating in the 30–65 W range. In summary, the SIMO inverters possess the key attributes of being (i) compact, (ii) low-cost, (iii) highly efficient, (iv) reliable, and (v) scalable to give many outputs, having (vi) good dynamic performance, (vii) no cross-coupling, (vii) low harmonic distortion, and (ix) higher power ratings, and providing for each output an independent and variable (x) voltage, frequency, and power flow.

REFERENCES

1. T. Li, "Single inductor multiple-output boost regulator," U.S. Patent 5757174, Jun. 13, 2000.
2. D. Ma, W.-H. Ki, P. K. T. Mok, and C.-Y. Tsui, "Single-inductor multiple-output switching converters with bipolar outputs", *Proc. IEEE Int. Symp. Circuits Syst.*, vol. 3, pp. 301–304, May 2001.
3. D. Ma, W.-H. Ki, C.-Y. Tsui, and P. K. T. Mok, "A 1.8 V single-inductor dual-output switching converter for power reduction technique", *Digest of Technical Papers IEEE VLSI Symposium Circuits*, pp. 137–140, Jun. 2001.
4. W.-H. Ki and D. Ma, "Single-inductor multiple-output switching converters", *Proc. IEEE PESC*, vol. 1, pp. 226–231, Jun. 2001.

5. D. Ma, W.-H. Ki, and C.-Y. Tsui, "A pseudo-CCM/DCM SIMO switching converter with freewheeling switching", *Proceedings of IEEE International Solid-State Circuits Conference*, vol. 1, pp. 1–3, Feb. 2002.

6. D. Ma, W.-H. Ki, C.-Y. Tsui, and P. K. T. Mok, "Single-inductor multiple-output switching converters with time- multiplexing control in discontinuous conduction mode", *IEEE J. Solid-State Circuits*, vol. 38, no. 1, pp. 89–100, Jan. 2003.

7. D. Ma, W.-H. Ki, and C.-Y. Tsui, "A pseudo-CCM/DCM SIMO switching converter with freewheel switching", *IEEE J. Solid-State Circuits*, vol. 38, no. 6, pp. 1007–1014, Jun. 2003.

8. C.-W. Leng, C.-H. Yang, and C.-H. Tsai, "Digital PWM controller for SIDO switching converter with time-multiplexing scheme", *International Symposium on VLSI Design Automation and Test*, pp. 52–55, Apr. 2009.

9. D. Kwon and G. A. Rincon-Mora, "Single-inductor-multiple-output switching DC-DC converters," *IEEE Trans. Circuits Syst., II: Exp. Briefs*, vol. 56, no. 8, pp. 614–618, Aug. 2009.

10. H. Eachempatti, S. Ganta, J. Silva-Martinez, and H. Martinez-Garcia, "SIDO buck converter with independent outputs", *2010 53rd IEEE International Midwest Symposium on Circuits and Systems (MWSCAS)*, pp. 37–40, Aug. 2010.

11. X. Jing, P. Mok, and M. Lee, "A wide-load-range constant-charge-auto-hopping control single-inductor- dual-output boost regulator with minimized cross-regulation", *IEEE J. Solid-State Circuits*, vol. 46, no. 10, pp. 2350–2362, Oct. 2011.

12. H. Chen, Y. Zhang, and D. Ma, "A SIMO parallel-string driver IC for Dimmable LED backlighting with local bus voltage optimization and single time-shared regulation loop", *IEEE Trans. Power Electron.*, vol. 27, no. 1, pp. 452–462, Jan. 2012.

13. Y.-J. Moon, Y.-S. Roh, J.-C. Gong, and C. Yoo, "Load-independent current control technique of a single-inductor multiple-output switching DC-DC converter", *IEEE Trans. Circuits Syst., II: Exp. Briefs*, vol. 59, no. 1, pp. 50–54, Jan. 2012.

14. S. Huynh and C. V. Pham, "Single inductor multiple LED string driver," U.S. Patent 20120043912 A1, Feb. 23, 2012.

15. A. Lee, J. Sin, and P. Chan, "Scalability of quasi-hysteretic FSM-based digitally- controlled single-inductor dual-string buck LED driver to multiple string", *IEEE Trans. Power Electron.*, vol. 29, no. 1, pp. 501–513, Jan. 2014.

16. Y. Zhang and D. Ma, "A fast-response hybrid SIMO power converter with adaptive current compensation and minimized cross-regulation", *IEEE J. Solid-State Circuits*, vol. 49, no. 5, pp. 1242–1255, May 2014.

17. C.-W. Chen and A. Fayed, "A low-power dual-frequency SIMO buck converter topology with fully-integrated outputs and fast dynamic operation in 45 nm CMOS", *IEEE J. Solid- State Circuits*, vol. 50, no. 9, pp. 2161–2173, Sep. 2015.

18. K. Modepalli and L. Parsa, "A scalable N-color LED driver using single inductor multiple current output topology", *IEEE Trans. Power Electron.*, vol. 31, no. 5, pp. 3773–3783, May 2016.

19. H. Ma, J. S. Lai, Q. Feng, W. Yu, C. Zheng, and Z. Zhao, "A novel valley-fill SPEIC-derived power supply without electrolytic capacitor for LED lighting application", *IEEE Trans. Power Electron.*, vol. 27, no. 6, pp. 3057–3071, Jun. 2012.

20. Y. Hu and M. Jovanovic, "LED driver with self-adaptive drive voltage", *IEEE Trans. Power Electron.*, vol. 23, no. 6, pp. 3116–3125, Nov. 2008.

21. K. I. Hwu and Y. T. Yau, "Applying one-comparator counter-based sampling to current sharing control of multichannel LED strings", *IEEE Trans. Ind. Appl.*, vol. 47, no. 6, pp. 2413–2421, Nov. 2011.

22. C. Y. Wum T. F. Wu, J. R. Tasi, Y. M. Chen, and C. C. Chen, "Multi-string LED backlight driving system for LCD panels with color sequential display and area control", *IEEE Trans. Ind. Electron.*, vol. 55, no. 10, pp. 3791–3800, Oct. 2008.

23. Q. Hu and R. Zane, "LED driver circuit with series-input-connected converter cells operating in continuous conduction mode", *IEEE Trans. Power Electron.*, vol. 25, no. 3, pp. 574–582, Mar. 2010.

24. W. Chen and S. Y. R. Hui, "A dimmable light-emitting diode (LED) driver with mag-amp postregulators for multistring applications", *IEEE Trans. Power Electron.*, vol. 26, no. 6, pp. 1714–1722, Jun. 2011.

25. S. G. Parler, "Application Guide, Aluminum Electrolytic Capacitors" (2015). [Online] Available: www.cde.com/resources/technical-papers/reliability.pdf.

26. Y. Guo, S. Li, A. T. L. Lee, S.-C. Tan, C. K. Lee, and S. Y. R. Hui, "Single-stage AC/DC single-inductor multiple-output LED drivers", *IEEE Trans. Power Electron.*, vol. 31, no. 8, pp. 5837–5850, Aug. 2016.

27. R. Johari, J. V. Krogmeier, and D. J. Love, "Analysis and practical considerations in implementing multiple transmitters for wireless power transfer via coupled magnetic resonance", *IEEE Trans. Ind. Electron.*, vol. 61, no. 4, pp. 1774–1783, Apr. 2014.

28. M. Q. Nguyen, Y. Chou, D. Plesa, S. Rao, and J.-C. Chiao, "Multiple-inputs and multiple-outputs wireless power combining and delivering systems", *IEEE Trans. Power Electron.*, vol. 30, no. 11, pp. 6254–6263, Nov. 2015.

29. B. H. Waters, B. J. Mahoney, V. Ranganathan, and J. R. Smith, "Power delivery and leakage field control using an adaptive phase array wireless power system", *IEEE Trans. Power Electron.*, vol. 30, no. 11, pp. 6298–6309, Nov. 2015.

30. J. Burdio, F. Monterde, J. Garcia, L. Barragan, and A. Martinez, "A two-output series-resonant inverter for induction-heating cooking appliances", *IEEE Trans. Power Electron.*, vol. 20, no. 4, pp. 815–822, Jul. 2005.

31. A. H. Mohammadian, et al., "Wireless power transfer using multiple transmit antennas," U.S. Patent US8629650 B2, 2008.

32. J. Lee, et al., "Wireless power transmitter and wireless power transfer method thereof in many-to-one communication," U.S. Patent US9306401 B2, 2011.

33. W. Jin, A. T. L. Lee, S. Li, S.-C. Tan, and S. Y. R. Hui, "Low-power multi-channel wireless transmitter", *IEEE Trans. Power Electron.*, vol. 33, no. 6, pp. 5016–5028, Jun. 2018.
34. A. T. L. Lee, W. Jin, S.-C. Tan, and S. Y. R. Hui, "Buck-boost single-inductor multiple-output high-frequency inverters for medium-power wireless power transfer", *IEEE Trans. Power Electron.*, vol. 34, no. 4, pp. 3457–3473, Apr. 2019.
35. W. Jin, A. T. L. Lee, S.-C. Tan, and S. Y. R. Hui, "A gallium nitride (GaN)-based single-inductor multiple-output (SIMO) inverter with multi-frequency AC outputs", *IEEE Trans. Power Electron.*, vol. 34, no. 11, pp. 10856–10873, Nov. 2019.
36. W. Jin, A. T. L. Lee, S.-C. Tan, and S. Y. R. Hui, "Single-inductor multiple-output inverter with precise and independent output voltage regulation", *IEEE Trans. Power Electron.*, vol. 35, no. 10, pp. 11222–11234, Oct. 2020.
37. A. T. L. Lee, W. Jin, S.-C. Tan, and S. Y. R. Hui, "Single-inductor multiple-output (SIMO) buck hybrid converter for simultaneous wireless and wired power transfer", *IEEE Journal of Emerging and Selected Topics in Power Electronics*, (early access), Jun. 2020.
38. The Qi Wireless Power Transfer System Power Class 0 Specification, Part 4: Ref. Designs, *Wireless Power Consortium*, v.1.2.3, Feb. 2017.

CHAPTER 2

Single-Inductor Multiple-Output DC-DC Converters

2.1 INTRODUCTION

Since single-inductor multiple-output (SIMO) DC-DC converters have been widely reported in the literature [1–12], the main purpose of this chapter is to give a general overview of SIMO DC-DC converters and cover the key aspects without delving into too much detail. It aims to provide a comprehensive survey on the existing SIMO DC-DC converters in terms of circuit topology, control method, and design consideration. Interested readers are encouraged to refer to the cited references for a detailed description of a particular type of SIMO DC-DC converter.

Modern electronic devices and systems require small, lightweight, reliable, robust, and efficient power supplies. Initially, low-dropout (LDO) linear regulators seem to be a good choice because it is a simple and inexpensive way to produce a regulated DC output voltage. The reason why they are called linear regulators is because the transistor, which is used to control the load current, operates in the linear region rather than in the saturation or cutoff region. In essence, the transistor acts like a variable resistor. Linear power regulators, underpinned by the concept of a voltage or current divider, are simply inefficient in that the difference between the

DOI: 10.1201/9781003239833-2

input voltage and the regulated output voltage is dissipated as heat. Low efficiency is the main disadvantage for this type of regulators. Despite the fact that they can produce a high-quality, tightly regulated output voltage, the output voltage must be *smaller* than the input voltage, which limits their real applications. In practice, they are mostly used in battery-powered applications. To generate multiple regulated DC output voltage or current levels, one approach is to use multiple point-of-load (PoL) LDO regulators, which draw power from a single switching converter. Figure 2.1 depicts the power management architecture composed of multiple LDO regulators. Even though this architecture uses only one inductor for multiple output voltages, the LDO regulators suffer from high power losses, which substantially reduce the battery life. In contrast, Figure 2.2 shows a different power management architecture that uses multiple DC-DC switching converters. By definition, a switching converter or regulator, also known as a switch-mode power supply, uses semiconductor power switches (e.g., field effect transistors) to transform the incoming power supply into a pulsed voltage, which is then smoothed by the inductor and capacitor to provide the required output voltage. A DC-DC converter performs the power conversion from an unregulated DC input voltage V_{in} to a regulated DC output voltage V_{out}, which can be higher or lower than V_{in}, depending on the specific circuit topology. High efficiency is a big advantage of inductive switching converters. This is mainly attributed to the very small on-resistance of the power transistor (e.g., power MOSFET), which incurs

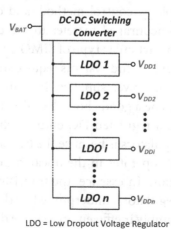

LDO = Low Dropout Voltage Regulator

FIGURE 2.1   Power management architecture based on multiple LDO regulators.

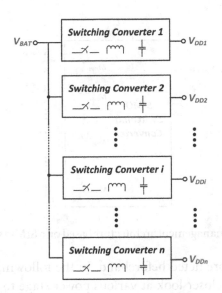

FIGURE 2.2 Power management architecture based on multiple DC-DC switching converters.

a small conduction loss when it is fully "ON". Also, when the transistor is in its "OFF" state, no current will flow across it. By operating it at a higher switching frequency, the values of the inductor and capacitors can be further reduced. Yet, inductors are still bulky and expensive in switching converters. To provide multiple DC output voltages, a simplistic solution is to use multiple independent DC-DC converters. But despite its high efficiency, the need for a large number of inductors significantly increases the size and cost of the overall power management unit or module, particularly for those applications requiring a greater number of output loads. Hence, SIMO DC-DC converters have emerged as a better alternative that leads to small form factor, low cost, and high efficiency. Figure 2.3 shows the power management architecture of a SIMO DC-DC converter.

A salient feature of SIMO converters is their extension (or scalability) to any arbitrary number of outputs. But in reality, only a finite number of outputs can be achieved in SIMO. This is conceivable because of the finite energy storage of the main inductor, which is largely determined by its inductance value and the maximum current rating. By using a SIMO DC-DC LED driver as a real example, generalized scalability models are formulated. Based on these analytical models, the maximum achievable number of outputs of a SIMO LED driver under different modes of

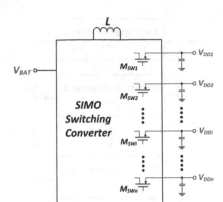

FIGURE 2.3   Power management architecture based on a SIMO switching converter.

operation can be predicted beforehand. In the following section, we will begin by taking a closer look at various power stage topologies of SIMO DC-DC converters.

2.2 POWER STAGE TOPOLOGIES

Topologically, SIMO converters can be treated as circuit extrapolations of single-inductor single-output (SISO) converters with the exception that the energy flow and control scheme becomes more complex. In Chapter 1, we have seen how two separate SISO buck converters can be combined into a single-inductor dual-output (SIDO) buck converter. In an SIDO buck converter, the inductor current is being multiplexed into two outputs. Similarly, we can extend this concept by multiplexing the inductor current of the SISO buck stage into N outputs in order to obtain a SIMO converter. The SIMO-based DC-DC converter employs only a single inductor in the power stage to produce multiple voltage (or current) levels. The inductor converts the DC input voltage into high-frequency pulsed DC current. Basically, it acts like a constant current source, which delivers the storage energy to each of the DC outputs sequentially in a round-robin time multiplexing (TM) manner. Each output contains a filter capacitor to mitigate the high-frequency switching ripples so as to produce a clean DC output voltage. The switching sequence of the power switches of the SIMO converter should be carefully designed in such a way that each output voltage (or current) can be independently regulated with minimum cross-regulation (or cross-interference). By definition, cross-regulation occurs when a change in one output causes an unexpected change in another output. In fact, a major challenge of SIMO converter is to meet the power demands of each individual output while at the same time mitigating or eliminating the undesirable effects of cross-regulation among outputs. In theory, multiple independent output voltages without cross-regulation can be attained by using the TM control scheme while ensuring that the SIMO converter operates only in discontinuous conduction mode (DCM) or pseudo-continuous conduction mode (PCCM). Due to the small dead times between successive current pulses distributed to the parallel output branches in DCM (or PCCM), the outputs are fully decoupled in the time domain. Similar to their SISO counterparts, the nonisolated SIMO DC-DC converters can be classified into four basic types: buck (step-up), boost (step-down), inverting buck-boost, and noninverting buck-boost.

2.2.1 SIMO DC-DC Buck Converter

To start with, let us consider the SISO buck converter. Figure 2.4 shows the power stage topologies of an asynchronous versus synchronous SISO buck converter. For simplicity, the buck converter is modeled using ideal circuit elements without any parasitic element. For instance, the actual

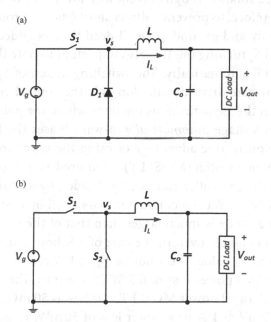

FIGURE 2.4 Power stage topology of (a) asynchronous SISO buck converter and (b) synchronous SISO buck converter.

power transistor is modeled as an ideal (nondissipative) switch. In a real circuit, it can be implemented as a power MOSFET.

In a typical asynchronous buck converter, there is only one switch (S_1) at the high side with a diode (D_1) at the low side. Both of them control the power to the load. When S_1 is turned ON, the input voltage charges the inductor (L) and output capacitor (C_o) and supplies the load current. When the output voltage reaches the desired value, S_1 is turned OFF, and the inductor voltage reverses its polarity, thus forward biasing the diode D_1. This allows the current to continue flowing through the inductor in the same direction. When a current flows across D_1, it is also known as being in freewheel mode. Hence, D_1 is also called a freewheeling diode. In the synchronous topology, the low-side D_1 is replaced by another switch S_2. To distinguish between the two switches, S_1 is called the high-side switch, and S_2 is the low-side switch. In steady-state condition, the gate drive signals for S_1 and S_2 are complementary to each other. This means that when S_1 is ON, S_2 must be OFF and vice versa. The on-off action of S_1 and S_2 must be precisely controlled because if both of them are turned on simultaneously even for a small period of time, a shoot-through current will flow from the input voltage to ground. Therefore, to prevent a direct short (shoot-through) between the power supply and ground, a small dead time is added between S_1 turning off and S_2 turning on. It is also important to note that S_2 *cannot* be switched OFF automatically. The switching action of S_2 needs to be provided by the gate drive circuit alongside the control circuitry. This is different than the asynchronous topology where the polarity reversal of the inductor voltage *automatically* forward biases the freewheeling diode D_1. Obviously, one advantage of using the asynchronous topology is that only one switch (MOSFET) is required as the control switch, which reduces the controller complexity. Besides, by default, there is no shoot-through issue. Yet the conduction loss attributed to the forward voltage (V_F) of a diode is much larger than that of the power MOSFET due to its on-resistance. Even in the case of a Schottky diode, which is known for a low V_F value, the value of V_F at 1 A can reach as high as 0.5 V, resulting in a power loss of 0.5 W. In contrast, the on-resistance of a typical N-channel power MOSFET is as low as 50 mΩ, which translates to only 50 mV at 1 A or a power loss of 50 mW. Hence, the use of MOSFET helps reduce losses significantly and thus increases the overall power efficiency.

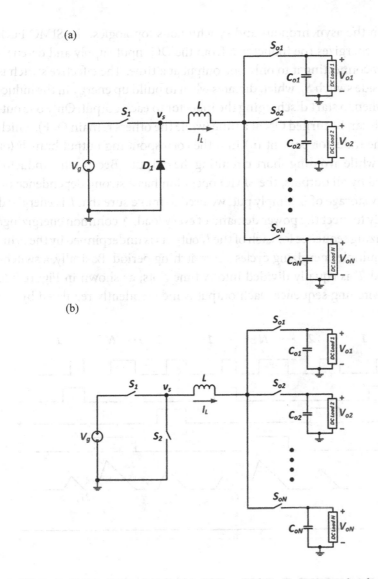

FIGURE 2.5 Power stage topology of (a) asynchronous SIMO buck converter
and (b) synchronous SIMO buck converter.

By TM the inductor current into several output paths, a SIMO DC-DC
buck converter with N outputs can be obtained. Figure 2.5 shows the
power stage topologies of an asynchronous versus a synchronous SIMO
buck converter. In general, the asynchronous SIMO buck converter
requires $(N+1)$ power switches, whereas its synchronous counterpart
requires $(N+2)$ power switches.

In both the asynchronous and synchronous topologies, the SIMO buck converter energizes the inductor L from the DC input supply and de-energizes L by connecting it to only one output at a time. The effective switch is the high-side switch S_1, which dictates when to build up energy in the inductor and when to start discharging the inductor to each output. Only one output switch can be turned ON at a time (while the others remain OFF), which diverts the inductor current (i_L) into the corresponding output branch (or channel) while avoiding short-circuiting the outputs. Because the inductor L is shared by all outputs, the SIMO operation has a strong dependence on the energy storage of L. Simply put, we need to make sure that L is energized sufficiently to meet the power demand of every load. A common energizing/de-energizing sequence for each of the N outputs is underpinned by the principle of multiple energizing cycles per switching period. Basically, a switching period T_s is equally divided into N time slots, as shown in Figure 2.6. In this switching sequence, each output is independently regulated by the

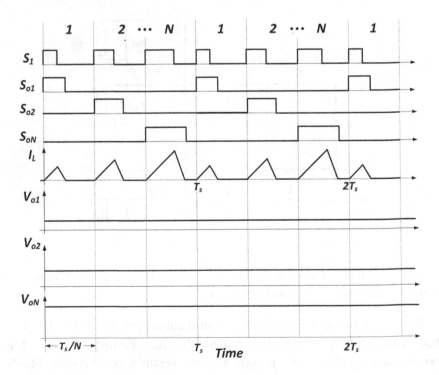

FIGURE 2.6 Theoretical waveforms of the gate-drive signals of the main switch and output switches, the inductor current, and the three output voltages of an asynchronous SIMO buck converter operating in DCM.

feedback controller to determine its duty ratio of the main switch or the output switch. DCM operation is preferred for SIMO buck converter because of two reasons. First, as in SISO buck operation, DCM transforms the complex-conjugate pair of poles into a dominant left-half plane pole, which simplifies the frequency compensation. Second, it enables the outputs to be fully decoupled in time, thus reducing the undesirable effects of cross-regulation. Figure 2.6 shows the theoretical waveforms of the gate drive signals of the main switch (S_1) and the output switches (S_{o1}, S_{o2}, ..., S_{oN}), the inductor current (I_L), and the output voltages (V_{o1}, V_{o2}, ..., V_{oN}) of an asynchronous SIMO buck converter operating in DCM.

Under steady-state condition, the average value of the voltage at the switching node $<v_s>$ is given by

$$\langle v_s \rangle = \frac{1}{N} V_{in} \sum_{k=1}^{N} D_{k1} \tag{2.1}$$

where N is the number of outputs, V_{in} is the input voltage, and D_{k1} is the on-time duty ratio of the high-side switch S_1.

Equation (2.1) physically represents how often the switching node v_s is connected to the input voltage V_{in} via the high-side switch S_1. Also, in DCM steady-state condition, the output voltage can be expressed as

$$V_{ok} = V_{in} \times \frac{D_{k1}}{D_{k1} + D_{k2}} \tag{2.2}$$

where V_{ok} is the DC output voltage at the kth output, D_{k1} is the on-time duty ratio of the high-side switch S_1, and D_{k2} is the on-time duty ratio of the low-side switch S_2 in synchronous buck converter (or the on-time duty ratio of the freewheeling diode D_1 in asynchronous buck converter). In particular, $V_{ok} < V_{in}$, for $D_{k1} > 0$ and $D_{k2} > 0$.

Equation (2.2) can be reexpressed as

$$V_{in} = V_{ok} \times \frac{D_{k1} + D_{k2}}{D_{k1}} \tag{2.3}$$

By substituting equation (2.3) into (2.1), we have

$$\langle v_s \rangle = \frac{1}{N} V_{in} \sum_{k=1}^{N} D_{k1} = \frac{1}{N} \sum_{k=1}^{N} V_{ok} (D_{k1} + D_{k2}) \tag{2.4}$$

Equation (2.4) is a general equation of $<v_s>$ in steady-state condition. It can be physically interpreted as how often the switching node v_s is connected to the input supply V_{in} via the high-side switch S_1. It also means how often v_s is connected to each individual output via the corresponding output switches. Notice that in steady state, the average voltage across the inductor L is zero, that is to say, the voltage at the left terminal of L is the same as that on the right (assuming a lossless inductor). Hence, the relationship in equation (2.4) holds true. In addition, this equation assumes *un-identical* duty ratios across all outputs. In other words, the SIMO converter produces unique DC output voltages (V_{ok}). This is often referred to as unbalanced load condition. For the special case of balanced load condition, the output voltages are the same (i.e., $D_{11}=D_{21}=...=D_{N1}=D_1$, $D_{12}=D_{22}=...=D_{N2}=D_2$, $V_{o1}=V_{o2}=...=V_{oN}=V_o$), and equation (2.4) can be reduced to the following form.

$$\langle v_s \rangle = D_1 V_{in} = (D_1 + D_2)V_o \tag{2.5}$$

2.2.2 SIMO DC-DC Boost Converter

A SISO DC-DC boost converter, shown in Figure 2.7, is another popular switched-mode power supply that produces a greater DC output voltage than the DC input voltage. By the same token, the SISO DC-DC boost converter can be extended to a SIMO DC-DC boost converter by multiplexing the inductor current into several outputs, as in Figure 2.8. It requires $(N+1)$ power switches.

Unlike the buck topology, the switching node v_s is always connected to the input voltage V_{in} via the inductor L in the boost topology. The closing of the low-side switch S_1 energizes L, whereas the closing of one of the output switches de-energizes L into the respective output. In steady state, the average voltage across L is zero. This implies that the average

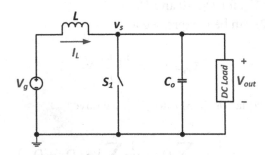

FIGURE 2.7   Power stage topology of a SISO DC-DC boost converter.

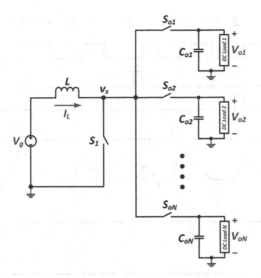

FIGURE 2.8   Power stage topology of a SIMO DC-DC boost converter.

voltage at the multiplexed output, i.e., the common node of all output branches, is equal to V_{in} by neglecting the DC resistance of L. To ensure a continuous flow of the inductor current, as soon as S_1 is switched OFF, one of the output switches must be switched ON immediately to complete the circuit. In addition, only one output switch can be switched ON at any point in time while the other switches must be switched OFF. It is worth mentioning that since the output voltage is greater than the input voltage in the boost topology, it is necessary to add a blocking diode at each output branch (either before or after the output switch) to prevent an unwanted reverse current flow from the output to the input supply. Figure 2.9 shows the switching sequence of the SIMO boost converter based on multiple energizing cycles per switching period. It includes the theoretical waveforms of the key signals such as the gate drive signals of the main switch (S_1) and the output switches (S_{o1}, S_{o2},..., S_{oN}) alongside the inductor current (I_L) and the output voltages (V_{o1}, V_{o2}, ..., V_{oN}).

Based on the steady-state operation of the SIMO boost converter, the average value of the voltage at the switching node $\langle v_s \rangle$ can be expressed as

$$\langle v_s \rangle = V_{in} = \frac{1}{N} \sum_{k=1}^{N} \frac{D_{k2} V_{ok}}{D_{k1} + D_{k2}} \qquad (2.6)$$

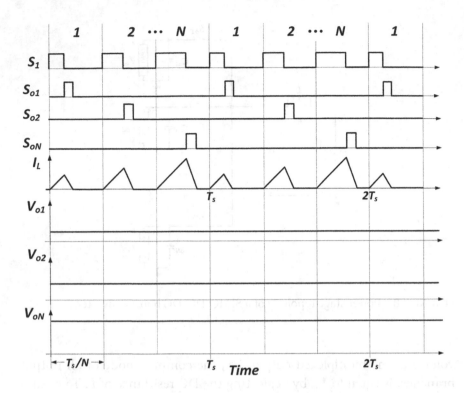

FIGURE 2.9 Theoretical waveforms of the gate-drive signals of the main switch and output switches, the inductor current, and the three output voltages of a SIMO boost converter operating in DCM.

The physical meaning of equation (2.6) is that the switching node is connected to the input supply at all times. Besides, it represents how often each individual output is connected to the switching node and also to the input supply via L. Equation (2.6) also shows that V_{ok} is greater than V_{in} for $D_{k1} > 0$ and $D_{k2} > 0$. In particular, when the SIMO boost converter operates at the boundary condition between CCM and DCM, commonly known as boundary conduction mode (BCM) or critical conduction mode, the third subinterval in DCM (or idle phase) disappears. Mathematically, $D_{k1} + D_{k2} = 1$. Hence, equation (2.6) can be rewritten as

$$\langle v_s \rangle = V_{in} = \frac{1}{N} \sum_{k=1}^{N} D_{k2} V_{ok} \qquad (2.7)$$

In balanced load condition, the duty ratios and the output voltages are identical, i.e., $D_{11}=D_{21}=\dots=D_{N1}=D_1$, $D_{12}=D_{22}=\dots=D_{N2}=D_2$, $V_{o1}=V_{o2}=\dots=V_{oN}$, and equation (2.6) can be reduced to

$$\langle v_s \rangle = V_{in} = \frac{D_2 V_o}{D_1 + D_2} \tag{2.8}$$

In particular, in BCM, equation (2.8) can be further reduced to

$$\langle v_s \rangle = V_{in} = D_2 V_o = (1 - D_1) V_o \tag{2.9}$$

From equation (2.9), the voltage conversion ratio M in BCM is given by

$$M = \frac{V_o}{V_{in}} = \frac{1}{(1 - D_1)} \tag{2.10}$$

2.2.3 SIMO DC-DC Inverting Buck-Boost Converter

Inverting buck-boost converter is considered the mirror of the boost counterpart because the positions of the main switch S_1 and the inductor L are swapped. Figure 2.10 shows the power stage topology of the SIMO inverting buck-boost converter. This converter requires $(N+1)$ power switches.

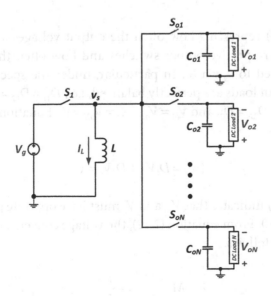

FIGURE 2.10 Power stage topology of a SIMO DC-DC inverting buck-boost converter.

In this topology, L is connected between the switching node v_s and ground on the low side, whereas in the boost topology, L is connected between V_{in} and v_s on the high side. Hence, the SIMO power train of the inverting buck-boost converter is opposite to that of the boost converter, albeit analogous circuit operations. Similar to the boost converter, S_1 is used to energize L, and the output switches (S_{o1}, S_{o2},..., S_{oN}) are used to de-energize L into their corresponding outputs. However, the inductor current and the output current now flow in the opposite direction. The inductor current flows from v_s to ground, and the output current flows from the output to v_s via the respective output switch. As a result, the polarity of the output voltages will be opposite to that of V_{in}, which explains why this circuit is called an "inverting" buck-boost converter. The theoretical waveforms of the gate-drive signals of the main switch (S_1) and the output switches (S_{o1}, S_{o2}, ..., S_{oN}) and the inductor current (I_L) are the same as those of the boost converter. In steady state, the average voltage across the inductor L is zero, which implies that <v_s> is also zero. By definition, <v_s> is equal to the weighted average of the input and output voltages. Hence, we can write

$$\langle v_s \rangle = \frac{1}{N} \left(V_{in} \sum_{k=1}^{N} D_{k1} + \sum_{k=1}^{N} D_{k2} V_{ok} \right) = 0 \qquad (2.11)$$

Equation (2.11) represents how often the output voltages are connected to v_s via their respective output switches and how often the input voltage is connected to v_s via S_1. In particular, under the special condition when the output loads are perfectly balanced, i.e., $D_{11} = D_{12} = ... = D_{1N} = D_1$, $D_{12} = D_{22} = ... = D_{2N} = D_2$, and $V_{o1} = V_{o2} = ... = V_{oN} = V_o$. Equation (2.11) can be reduced to

$$\langle v_s \rangle = D_1 V_{in} + D_2 V_o = 0 \qquad (2.12)$$

Equation (2.12) indicates that V_{in} and V_o must have opposite polarity since $D_1 > 0$ and $D_2 > 0$. From equation (2.12), the voltage conversion ratio M can be obtained as follows.

$$M = \frac{V_o}{V_{in}} = -\frac{D_1}{D_2} \qquad (2.13)$$

When the converter operates in BCM, $D_2 = (1 - D_1)$. Hence, equation (2.13) becomes

$$M = \frac{V_o}{V_{in}} = -\frac{D_1}{(1 - D_1)} \tag{2.14}$$

2.2.4 SIMO DC-DC Noninverting Buck-Boost Converter

Basically, a SISO noninverting buck-boost topology is derived by simply cascading the SISO buck and boost stages. Likewise, a SIMO noninverting buck-boost converter is formulated by cascading the SISO buck input stage and the SIMO boost output stage. As the name implies, "noninverting" means the output voltages have the *same* polarity as the input voltage. Figure 2.11 contains the power stage topology of the SIMO noninverting buck-boost converter. The total number of power switches required in this topology is $(N+3)$, compared to $(N+1)$ in its inverting counterpart.

In this topology, the converter can be configured as a noninverting buck converter by opening (disabling) the low-side switch S_3 at all times. In this case, it behaves like a SIMO buck converter. The high-side switch S_1 and one of the output switches (S_{o1}, S_{o2}, ..., S_{oN}) energize the inductor L.

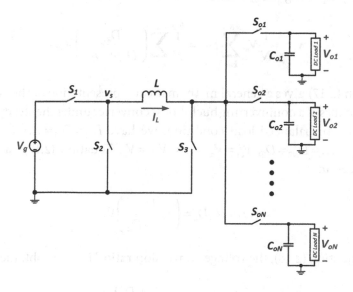

FIGURE 2.11   Power stage topology of a SIMO DC-DC noninverting buck-boost converter.

The low-side switch S_2 and one of the output switches de-energize L. The average value of the switching node $\langle v_s \rangle$ is given by

$$\langle v_s \rangle = \frac{1}{N} V_{in} \sum_{k=1}^{N} D_{k1} \tag{2.15}$$

On the other hand, the converter can also be configured as noninverting boost converter by closing (enabling) the high-side switch S_1 at all times. Thus, it acts like a SIMO boost converter. The low-side switch S_3 energizes L, while one of the output switches de-energizes L. Consider the voltage at the common node (v_c) to which all output branches are connected. The average value of this common node $\langle v_c \rangle$ is virtually the same as that of the switching node of a SIMO boost converter, which is given by equation (2.6). Hence, we can write

$$\langle v_c \rangle = \frac{1}{N} \sum_{k=1}^{N} \frac{D_{k2} V_{ok}}{D_{k1} + D_{k2}} \tag{2.16}$$

From Figure 2.11, it can be observed that the difference between v_s and v_c is just the voltage across the inductor L. In steady state, the average voltage across L is zero, and therefore, $\langle v_s \rangle = \langle v_c \rangle$, assuming a lossless inductor. Hence, by equating (2.15) and (2.16), we have

$$\langle v_s \rangle = \frac{1}{N} V_{in} \sum_{k=1}^{N} D_{k1} = \frac{1}{N} \sum_{k=1}^{N} \left(\frac{D_{k2}}{D_{k1} + D_{k2}} \right) V_{ok} \tag{2.17}$$

Equation (2.17) gives a general mathematical representation of the switching behaviors of a noninverting buck-boost converter under the steady-state condition. In balanced load condition, we have $D_{11} = D_{21} = \ldots = D_{N1} = D_1$, $D_{12} = D_{22} = \ldots = D_{N2} = D_2$, $V_{o1} = V_{o2} = \ldots = V_{oN} = V_o$. Equation (2.17) can thus be reduced to

$$\langle v_s \rangle = V_{in} D_1 = \left(\frac{D_2}{D_1 + D_2} \right) V_o \tag{2.18}$$

From equation (2.18), the voltage conversion ratio M can be obtained as

$$M = \frac{V_o}{V_{in}} = \frac{D_1 (D_1 + D_2)}{D_2} \tag{2.19}$$

In particular, when the noninverting buck-boost converter operates in BCM, the conversion ratio M becomes

$$M = \frac{V_o}{V_{in}} = \frac{D_1}{(1 - D_1)} \tag{2.20}$$

2.3 CONTROL SCHEME

A DC-DC converter, regardless of SISO or SIMO, consists of two major functional blocks: power stage and controller. The power stage uses the switching elements to transform the input voltage to the desired output. The controller supervises the switching sequence of the power switches and regulates the output voltage (or current) against load and line transients as well as any variation in component parameters and ambient condition. The latter is key to the stability and precision of a power supply. The two are linked by a negative feedback loop that compares the sensed output to the desired output to derive the necessary control signals. In general, there are two distinct control schemes for switching power supplies, namely, hysteretic control and pulse-width modulation (PWM) control [3,7,13–19].

Hysteretic voltage (or current) mode control is prominent for its inherent (unconditional) stability [15,20]. The simplicity of hysteretic DC-DC converters makes them superior to their PWM counterparts in terms of size and design complexity. Yet, hysteretic voltage mode control only works for buck converters with output capacitors that have relatively large equivalent series resistance (ESR). To ensure normal operation, the ESR must be sufficiently large so that the variation of the output voltage closely mimics that of the inductor current. Unfortunately, many emerging applications demand ceramic or tantalum capacitors with low ESR to achieve low noise, small ripples, and increased efficiency. Besides, the output of either boost or buck-boost topology carries no information about the inductor current ripple. Hence, a current sensor is needed in order to explicitly sense the inductor current, which increases the design complexity. Alternatively, hysteretic current mode control can also be used. It senses the inductor current and keeps it between preset upper and lower boundaries. Thus, the average inductor current is directly regulated. Nonetheless, a major issue with the conventional hysteretic control is its variable switching frequency, which leads to the injection of high-frequency harmonics into the system and increased switching losses. To further complicate matters, such harmonics are load-dependent, which are often difficult to predict and filter. Subsequently, fixed-frequency hysteretic control schemes with simple feedback network are proposed for switching converter [7,14], which carry the key advantages of ease of implementation, relatively low harmonics, limited electromagnetic interference, and high efficiency.

Another popular control scheme is the conventional PWM control with fixed frequency. Fundamentally, there are two ways of generating the PWM signals: voltage-mode control and current-mode control. In PWM voltage-mode control, a control voltage (or "error voltage") is generated by the error amplifier (EA) by subtracting the feedback voltage from the reference voltage. The PWM signal is then generated by comparing the control voltage with an external sawtooth voltage (or "PWM ramp") running at a fixed switching frequency. PWM voltage-mode control, albeit simple and easy to implement due to the use of a single feedback loop, suffers from slow load or line transient response. Loop compensation becomes more challenging due to the LC output filter and the fact that the loop gain varies with the input voltage. To attain a fast and stable transient response, a fairly complicated Type III compensation is required. To address the deficiencies of the voltage-mode control method, a current-mode control method was introduced in the early 1980s. A voltage-mode-controlled converter can be transformed into a current-mode one by adding an inner inductor current sensing path and feedback loop. In current-mode control, the PWM signal is generated by comparing the control voltage with the sensed inductor current derived from the inner (second) loop. Conceptually, the inner current loop makes the inductor a voltage-controlled current source, effectively eliminating it from the outer voltage control loop at DC and low frequency. Therefore, the power stage with closed current loop becomes a *first-order* system, as opposed to a second-order system with LC resonance under voltage-mode control. This reduces the phase lag caused by the power stage poles from 180° to 90°, which makes it much easier to compensate the output voltage loop. The key is to ensure that there is sufficient phase and gain margins at the unity-gain frequency. In general, the current-mode-controlled circuit exhibits a higher-gain bandwidth, which translates into high precision and faster response, compared to the voltage-mode-controlled circuit. The discussion of loop compensation design for PWM control is beyond the scope of this book. Interested readers are recommended to refer to the relevant literature for more details.

The hysteretic or PWM control scheme for a SISO converter generally applies to a SIMO converter with the exception that some modifications are needed to allow independent control over multiple outputs. Regardless of the underlying control scheme, the challenge in SIMO power supplies is to combine multiple feedback paths to generate the switch-control signals at appropriate times for achieving precise and independent output regulation. Chapter 1 presents the two main switching schemes of SIMO converters, i.e., the multiple energizing method and single energizing method. In the following subsections, let us look at how these two switching methods can actually be realized in either hysteretic or PWM control scheme.

2.3.1 Multiple-Energizing Method

The main objective of this method is to energize the inductor multiple times within a full operating period T. To fully decouple the outputs in time, T is equally divided into N subintervals, where N is the number of outputs. Each of the N time slots will be dedicated to controlling only a single output. From a practical standpoint, an N-to-1 time multiplexer is needed in the control logic to enable time-division multiplexing of N feedback loops. The select signals of this multiplexer, generated by an N-phase nonoverlapping clock generator running at the switching frequency, will choose the output of a particular hysteretic comparator to be connected to the subsequent gate drive circuits, hence closing one feedback loop in each T/N subinterval. Figure 2.12a shows the logic diagram of the hysteretic

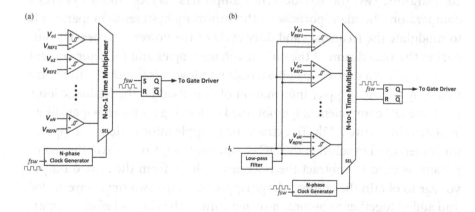

FIGURE 2.12 Logic diagram of TM (a) hysteretic voltage-mode control and (b) hysteretic current-mode control of SIMO buck converter.

voltage-mode control scheme of a SIMO buck converter. In voltage-mode control, each individual output voltage (or its scaled-down version derived from a voltage divider) is fed back to the corresponding hysteretic comparator, where it is compared against a fixed reference voltage. For example, the first output voltage V_{o1} will be compared with its reference voltage V_{REF1}. The output voltage rises with the inductor current i_L as the inductor L energizes (charges). As soon as it becomes greater than the reference voltage (i.e., the upper hysteresis limit), L de-energizes (discharges), and the output voltage begins to drop. As mentioned before, the feedback loops are stable as long as the ESR of each output capacitor is large enough to induce i_L-like ripples in the output voltages. That means by feeding back and mixing the outputs, we are essentially sensing both the output voltages and i_L.

Unlike buck converters, the inductor current ripple Δi_L in boost or buck-boost converters does *not* completely flow toward the output. Even for buck converters, the use of low-ESR output capacitor can also result in a lack of Δi_L information in the output. In such cases, the output voltage alone does *not* reflect the AC ripples of the inductor current. Hence, the inductor current i_L must also be sensed and combined with the output error voltage in the feedback loop to achieve current-mode-like control. Figure 2.12b shows the logic diagram of the hysteretic current-mode control of SIMO buck converter. Notice that the schematic symbol of a summing hysteresis comparator in Figure 2.12b is intended to give a high-level abstraction. In actual implementation, it can be made up of two comparators, two transconductance amplifiers (OTAs) and a hysteresis comparator. The main purpose of the summing hysteresis comparator is to modulate the frequency and duty cycle of the power switches by comparing the scaled sum of the output voltage ripples and inductor current ripples against a user-defined hysteresis window. It is important to realize that only the AC (ripple) information of i_L is used by the feedback loop. First, the DC component of i_L is obtained by feeding i_L into a low-pass filter, as shown in Figure 2.12b. To extract the i_L ripple information, a comparator is then used to subtract the DC component of i_L from i_L. Another comparator is used to subtract the reference voltage from the sensed output voltage to obtain the output voltage ripple. Then the two ripples are scaled and added together to become a scaled sum of the ripples before comparing it with a reference value (e.g., 0 V). Finally, an N-to-1 multiplexer is used to combine the outputs of the hysteresis comparators and select only one of them for controlling the switches in every T/N subinterval.

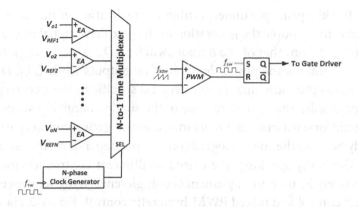

FIGURE 2.13 Logic diagram of TM PWM voltage-mode control of SIMO buck converter.

Based on time division multiplexing, the PWM control scheme for SIMO converter can be derived by extrapolating from that of the SISO counterpart. Figure 2.13 shows the logic diagram of the PWM voltage-mode control scheme of SIMO buck converter. Each EA feeds its output to a N-to-1 multiplexer, which selects one of the EAs and then connects it to the subsequent PWM comparator. The PWM comparator compares the slow-moving analog output signal of EA against an external sawtooth signal to produce a PWM signal for controlling the main power switch. Under the trailing-edge modulation, the main switch will be turned ON at the rising edge of a master (system) clock, and the inductor L will be energized. When the modulating ramp of the sawtooth signal reaches the level of the EA's output voltage, the main switch will be turned OFF, and L will be de-energized. The effective duty cycle of the main switch using trailing-edge modulation is determined by the ON time of the switch. Other modulation schemes such as leading-edge modulation and double-edge modulation can also be employed. Finally, PWM voltage control can be transformed into PWM current control by replacing the sawtooth waveform with the sensed inductor current from the inner current loop.

2.3.2 Single-Energizing Method

Instead of energizing the inductor multiple times within a full operating cycle, another approach is to energize the inductor (L) only once per cycle, but with sufficient energy to supply all loads. During the de-energizing phase, the energy from the inductor will be released successively to each output. The key principle underpinning the realization of this energizing sequence is to decouple the generation of the gate drive signal of the main switch (D_{main}) from that of the output switches (D_{o1}, D_{o2}, ..., D_{oN}). In this way, D_{main} contains the power demand of *all* outputs while (D_{o1}, D_{o2}, ..., D_{oN}) contains the individual power demand specific to their corresponding outputs only. The control scheme of the single-energizing sequence is less straightforward than that of its multiple-energizing counterpart. This is mainly because the control logic of one output depends on the preceding output. Generally speaking, there are two different control schemes that can effectively be used to implement the single-energizing sequence: pure hysteretic control and mixed PWM-hysteretic control. Figure 2.14a shows the logic diagram of the pure hysteretic current-mode control scheme of SIMO buck converter.

Firstly, let us consider the logic using the top SR latch for generating D_{main}, as shown in Figure 2.14a. The rising edge of the master clock f_{sw} sets the SR latch of D_{main}, which triggers the energizing event of the SIMO converter, where f_{sw} is the same as the switching frequency. Since the initial error of the last output voltage V_{oN} indicates whether or not the inductor L has adequate energy to supply all outputs, this error sets how long L needs to be energized. This error is modulated by the AC ripple of the inductor current i_L to become the error voltage v_{error}. When v_{error} equals one, it will reset the SR latch of D_{main} and set the SR latch of D_1, which initiates the de-energizing phase by discharging L to the first output V_{o1}. When V_{o1} becomes greater than its reference voltage V_{REF1}, the SR latch of

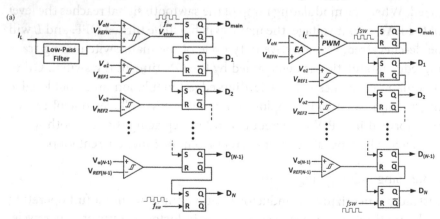

FIGURE 2.14 Logic diagram of (a) pure hysteretic current-mode control and (b) hybrid PWM-hysteretic control of SIMO buck converter.

D_1 will be reset, and the SR latch of D_2 will be set. The same sequence will repeat until the last (N^{th}) output is reached. But this time around, it is the master clock which resets the SR latch of D_N and then sets the SR latch of D_{main} for the next cycle. On the other hand, Figure 2.14b illustrates the logic diagram of the hybrid PWM-hysteretic control method. An EA is used to generate a compensating error voltage of V_{oN}, which acts as the peak threshold of the inductor current. The inductor current is compared with this peak threshold to determine the end of the energizing event and the onset of the de-energizing event. The remaining control logic is similar to that of the pure hysteretic control. Nonetheless, there is a major drawback of using only the error of the last output to determine the energizing interval. Imagine when one or more outputs suddenly draw more energy away from the inductor. It is quite likely that the energy left in the inductor will be insufficient to supply not only the last output but also the last few outputs. To resolve this issue, some researchers proposed combining the errors of *all* outputs rather than exploiting the cumulative effect on the last output [21,22].

2.4 SCALABILITY MODEL

A niche application of the SIMO DC-DC converter is to serve as a multistring LED driver for general lighting or color mixing applications. For example, a single-inductor three-output (SITO) DC-DC converter can be configured as an SITO LED driver that can supply three independent LED strings comprising of red, green, and blue LEDs, respectively. Naturally, we would like to find out the limits of extending it to SIMO with N independently driven LED strings. In this section, the theoretical maximum number of LED strings N_{max} is determined in the SIMO-based architecture. Figure 2.15 shows the circuit diagram of an ideal SIMO buck-type LED driver consisting of N number of LED strings. Figure 2.16 shows the timing diagram of the inductor current I_L, the gate drive signals of the two power switches (S_1, S_2), and the gate drive signals of the four output switches (S_a, S_b, S_c, S_d) of a SIMO buck LED driver. To simplify the ensuing analysis, balanced load condition is assumed, which means that the output voltage and current are the same across all strings. Based on the multiple-energizing method, energy stored in the main inductor L is transferred to each individual output *only once* within a total of N switching phases.

For illustration purpose, let us refer to Figure 2.16. In a particular output, the output switch is turned OFF during $D_{3n}T_s$, where D_{3n} is the duty ratio of the third subinterval (idle phase) of DCM, the subscript n denotes the output index, and T_s is the switching period. In this subinterval, the output capacitor supplies energy to the LED string. During the subsequent $(N-1)$ cycles, the output switch remains OFF, and the output capacitor continues to discharge to the respective LED string. Hence, the total discharging time of the output capacitor t_{dch} can be expressed as

$$t_{dch} = D_3 T_s + (N-1)T_s = (D_3 + N - 1)T_s \tag{2.21}$$

During the discharging interval, the output capacitor is connected to the LED string that acts like a constant current sink. Assume the output capacitor is ideal with no ESR. The output voltage is effectively the same as the voltage across the output capacitor $v_c(t)$ that can be expressed as the charge $q(t)$ divided by the capacitance value C_o, that is,

$$v_c(t) = \frac{q(t)}{C_o} = \frac{1}{C_o}\int_0^{t_{dch}} i_c(\tau)d\tau + v_c(0) \tag{2.22}$$

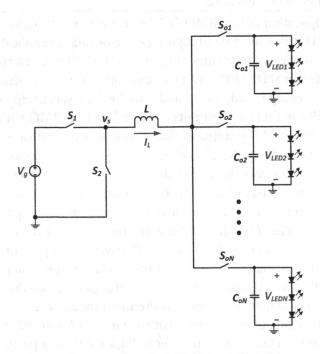

FIGURE 2.15   Circuit diagram of an ideal SIMO buck LED driver.

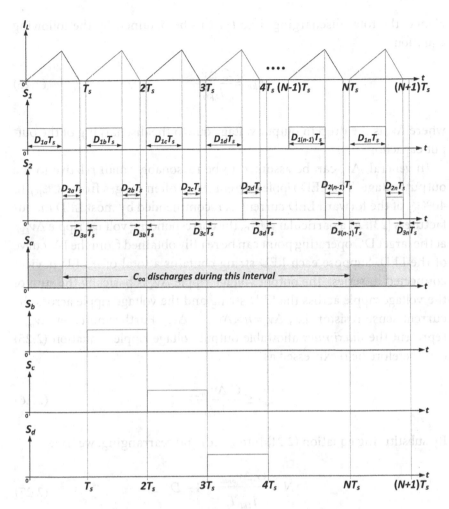

FIGURE 2.16 Timing diagram of the inductor current I_L, the two power switches (S_1, S_2), and the first four output switches (S_a, S_b, S_c, S_d) of a SIMO buck LED driver.

Since the output capacitor is connected to the LED string in series, the capacitor current is the same as the LED current. Hence, we can write

$$i_c(\tau) = I_{LED} \tag{2.23}$$

By substituting equation (2.23) into (2.22), we have

$$\Delta v_o = \Delta v_c = v_c(t) - v_c(0) = \frac{1}{C_o}\left(I_{LED}t_{dch}\right) \tag{2.24}$$

Hence, the total discharging time t_{dch} can be obtained by the following equation.

$$t_{dch} = \frac{C_o \Delta v_o}{I_{LED}}$$ (2.25)

where Δv_o is the drop in output voltage due to the discharging of the output capacitor.

In general, Δv_o can be assumed to be reasonably small relative to the output voltage. The LED ripple current Δi_{LED} often ranges from $10\%_{P\text{-}P}$ to $40\%_{P\text{-}P}$ of the forward LED current as recommended by most LED manufacturers [23]. At a particular Δi_{LED}, the corresponding voltage ripple Δv_{LED} at the target DC operating point can be readily obtained from the I-V curve of the LED. Suppose each LED string contains a total of n LED devices connected in series. The output voltage ripple Δv_o is basically the sum of the voltage ripple across the LED string and the voltage ripple across the current sense resistor, i.e., $\Delta v_o = n \times \Delta v_{LED} + \Delta v_{cs}$. Furthermore, let Δv_{omax} represent the *maximum* allowable output voltage ripple. Equation (2.25) can therefore be reexpressed as

$$t_{dch} \leq \frac{C_o \Delta v_{omax}}{I_{LED}}$$ (2.26)

By substituting equation (2.21) into (2.26) and rearranging, we have

$$N \leq \frac{C_o \Delta v_{omax}}{I_{LED} T_s} + 1 - D_3$$ (2.27)

Hence, the theoretical maximum number of LED devices per string (N_{max}) is given by

$$N_{max} = \frac{C_o \Delta v_{omax}}{I_{LED} T_s} + 1 - D_3 = \frac{C_o \Delta v_{omax}}{I_{LED} T_s} + D_1 + D_2$$ (2.28)

Since N_{max} is obviously an integer value, the *floor* function is used to round the result down to the closest integer. Hence, equation (2.28) becomes

$$N_{max} = floor\left(\frac{C_o \Delta v_{omax} + I_{LED} T_s (1 - D_3)}{I_{LED} T_s} \right)$$ (2.29)

Equation (2.29) is a general equation which can be used to determine the scalability limit of a SIMO buck LED driver operating in DCM. For ease of discussion, it will be referred to as a DCM scalability model. In particular, when $D_3=0$, the SIMO driver will operate in BCM as the third subinterval ceases to exist. Equations (2.28) and (2.29) thus become equations (2.30) and (2.31), respectively.

$$N_{max} = \frac{C_o \Delta v_{o\,max}}{I_{LED} T_s} + 1 \tag{2.30}$$

$$N_{max} = floor\left(\frac{C_o \Delta v_{o\,max}}{I_{LED} T_s} + 1 \right) \tag{2.31}$$

The average inductor current of a single-output buck converter is identical to the DC output current. For a SIMO LED driver, the average inductor current I_{Lavg} is the sum of the individual LED currents I_{LED} in each string. Under the assumption of a balanced load condition, $I_{Lavg}=N \times I_{LED}$, where N is the total number of LED strings. In particular, I_{Lavg} attains its maximum value in BCM, resulting in maximum output (LED) current and thus, maximum output power [1]. In theory, the maximum achievable number of LED strings in SIMO LED driver can be written as

$$N_{max} = \frac{I_{Lavg,max}}{I_{LED}} \tag{2.32}$$

By simple geometry, $I_{Lavg,\,max}$ is given by the following equation [24].

$$I_{L\_avg\_max} = \frac{m_1 D_1 T_s}{2} = \frac{(V_g - V_o) D_1 T_s}{2L} \tag{2.33}$$

where V_g is the input voltage, and V_o is the output voltage.

By substituting equation (2.33) into (2.32) and rearranging, T_s can be obtained as follows.

$$T_s = \frac{2L N_{max} I_{LED}}{D_1(V_g - V_o)} \tag{2.34}$$

Now, by substituting equation (2.34) into (2.30) and rearranging, we have

$$2LI_{LED}{}^2 N_{max}{}^2 - 2LI_{LED}{}^2 N_{max} - C_o \Delta v_{omax}(V_g - V_o)D_1 = 0 \tag{2.35}$$

Equation (2.35) is a quadratic equation of N_{max}. The discriminant Δ of equation (2.35) can be written as

$$\Delta = 4L^2 I_{LED}{}^4 + 8LI_{LED}{}^2 C_o \Delta v_{omax}(V_g - V_o)D_1 > 0 \qquad (2.36)$$

Since $(V_g - V_o) > 0$ for a buck switcher, the discriminant in equation (2.36) has a positive value which implies that equation (2.35) has the two real roots (r_1, r_2) as follows.

$$r_1, r_2 = \frac{2LI_{LED}{}^2 \pm \sqrt{\Delta}}{4LI_{LED}{}^2} \qquad (2.37)$$

Because N_{max} must be a positive integer, the negative root is therefore eliminated, leaving only the positive root, i.e.,

$$N_{max,BCM} = floor\left(\frac{1}{2} \times \left[1 + \sqrt{1 + \frac{2C_o \Delta v_{omax} V_o (V_g - V_o)}{LI_{LED}{}^2 V_g}} \right] \right) \qquad (2.38)$$

Equation (2.38) defines the theoretical maximum total number of outputs in SIMO operating in BCM. It is referred to as a BCM scalability model, which is a special case of the DCM scalability model. As a matter of fact, it is interesting to observe that equation (2.32) also holds for the case of DCM. By simple geometry, the switching period T_s in DCM can be obtained as

$$T_s = \frac{2LN_{max} I_{LED}}{D_1(D_1 + D_2)(V_g - V_o)} \qquad (2.39)$$

In DCM, the duty ratio of the first subinterval D_1 can be expressed as

$$D_1 = M\sqrt{\frac{K}{1 - M}} \qquad (2.40)$$

where $M = \dfrac{V_o}{V_g}$ and $K = \dfrac{2L}{R_L T_s} = \dfrac{2LI_{LED}}{V_o T_s}$ [24].

Realizing that the derivations that lead to equation (2.38) can also be applied to the case of DCM. The theoretical maximum total number of

LED strings of a SIMO converter operating in DCM can therefore be written as

$$N_{max,DCM} = floor\left(\frac{1}{2} \times (1 - D_3) \times \left[1 + \sqrt{1 + \frac{2C_o \Delta v_{omax}(V_g - V_o)D_1}{LI_{LED}^2(1 - D_3)}}\right]\right) \quad (2.41)$$

Notice that in BCM, $D_3 = 0$ and $D_1 = V_o/V_g$, and equation (2.41) thus reduces to equation (2.38). Basically, equation (2.41) is a general equation that can be used to obtain the theoretical maximum number of outputs (LED strings) in SIMO operating in either DCM or BCM. It is also worth noting that the average inductor current in DCM is smaller than that in BCM. As a consequence, at the same current value per string, the maximum number of outputs of SIMO in DCM is less than that in BCM, i.e., $N_{max, DCM} < N_{max, BCM}$.

From a practical standpoint, the ESR of the output capacitor R_{ESR} needs to be taken into consideration. Since the same current will flow through both C_o and R_{ESR}, it results in an additional voltage drop of $\Delta V_{ESR} = I_{LED} \times R_{ESR}$. Thus, Δv_o can be expressed as

$$\Delta v_o = \Delta v_c + \Delta v_{ESR} = \Delta v_c + I_{LED} \times R_{ESR} \quad (2.42)$$

Rearranging the terms in equation (2.42), we have

$$\Delta v_c = \Delta v_o - I_{LED} \times R_{ESR} \quad (2.43)$$

Hence, by taking into account the effect of R_{ESR}, equations (2.38) and (2.41) become equations (2.44) and (2.45), respectively.

$$BCM: N_{max\_BCM} = floor\left(\frac{1}{2} \times \left[1 + \sqrt{1 + \frac{2C_o V_o(\Delta v_{omax} - I_{LED}R_{ESR})(V_g - V_o)}{LI_{LED}^2 V_g}}\right]\right) \quad (2.44)$$

$$DCM: N_{max\_DCM} = floor\left(\frac{1}{2} \times (1 - D_3) \times \left[1 + \sqrt{1 + \frac{2C_o(\Delta v_{omax} - I_{LED}R_{ESR})(V_g - V_o)D_1}{LI_{LED}^2(1 - D_3)}}\right]\right)$$
$$(2.45)$$

The presence of R_{ESR} in equations (2.44) and (2.45) reduces the theoretical maximum achievable number of outputs in SIMO. Practically speaking, it is therefore recommended to choose an output capacitor with a smaller ESR, whenever possible. Figure 2.17 shows the theoretical maximum

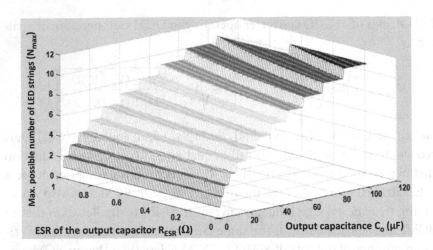

FIGURE 2.17   Theoretical maximum number of LED strings (N_{max}) versus the output capacitance (C_o) and the ESR of the output capacitor under the BCM scalability model.

number of LED strings versus the output capacitance and the ESR values under the BCM scalability model for an LED current of 80 mA per string and a maximum current ripple requirement of $40\%_{P\text{-}P}$.

Intuitively, at a particular LED current, an increasing number of LED strings can be achieved by using a larger output capacitor with the same ESR value. For instance, if the output capacitance is increased from 4.7 to 22 µF (while the ESR remains at 100 mΩ), the BCM model from equation (2.44) suggests that the maximum number of LED strings can be increased from three to six. It is also interesting to note that the maximum number of outputs has a stronger dependence on the value of the output capacitor than its ESR value, as shown in Figure 2.17.

On the other hand, it is crucial to investigate how the LED current value affects the maximum achievable number of LED strings in SIMO. For the sake of our discussion, let us assume that we have a SIDO LED driver with the following parameter values: $L = 47$ µH, $C_o = 4.7$ µF, $R_{ESR} = 100$ mΩ, $V_g = 15$ V, and $V_o = 6.4$ V. Also, each of the two LED strings consists of two series-connected white LED devices targeted for LCD backlighting applications. By using the BCM scalability model, the relationship between N_{max} and I_{LED} can easily be obtained across different values of maximum output voltage ripple $\Delta v_{o,\,max}$. Based upon the I-V curve given in the datasheet or the SPICE model of the chosen LED device, $\Delta v_{o,\,max}$ can be estimated from the LED current ripple Δi_{LED}. As an example, a $20\%_{P\text{-}P}$ current ripple corresponds to about

$2\%_{P\text{-}P}$ voltage ripple, while a $40\%_{P\text{-}P}$ current ripple corresponds to about $4\%_{P\text{-}P}$ voltage ripple. Figure 2.18 shows a plot of N_{max} versus I_{LED} for Δi_{LED} ranging from $5\%_{P\text{-}P}$ to $40\%_{P\text{-}P}$. This plot is beneficial to a practical design of SIMO LED driver in two ways. First, based on the BCM model, the theoretical maximum number of LED strings that can be achieved at a particular LED current and current ripple requirement can be obtained directly from the plot. Second, the maximum allowable LED current value for a SIMO with a fixed number of LED strings can also be obtained from the plot. For example, given a 20% current ripple requirement (i.e., $\Delta i_{LED}=20\%_{P\text{-}P}$), a SIDO (dual-string) configuration is viable as long as the LED current in each string does *not* exceed 100 mA. If an LED current greater than 100 mA is needed for a particular application, two options can be considered: (i) Relax the current ripple requirement, if possible. A wider tolerance in Δi_{LED} is generally acceptable since the ripple frequency of a DC-DC switching converter is too high for the human eyes to detect; (ii) operate the SIMO buck converter in PCCM [2]. Unlike DCM, the average value of the inductor current is increased in PCCM due to a positive DC offset.

Interestingly, the scalability model can be further extended to PCCM by adding a DC component to the average inductor current. By performing similar derivations as in DCM, the theoretical maximum number of outputs of SIMO operating in PCCM can be obtained as follows.

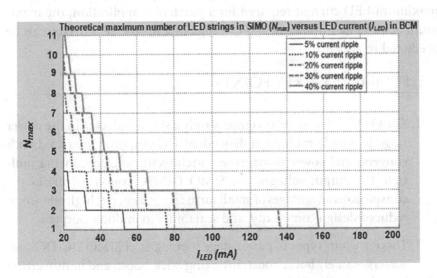

FIGURE 2.18 Relationship between the maximum achievable number of LED strings and the LED current per string under the BCM scalability model.

$$N_{\max,PCCM} = floor\left(\begin{array}{c} \dfrac{1}{2I_{LED}} \times \left[\left(I_{DC}+(1-D_3)I_{LED}\right)\right] \\ \times \left[1+\sqrt{1+\dfrac{2C_o(\Delta v_{o\max}-I_{LED}R_{ESR})(V_g-V_o)D_1(1-D_3)}{L\left[I_{DC}-(1-D_3)I_{LED}\right]^2}}\right] \end{array}\right)$$

$$(2.46)$$

It is important to realize that equation (2.46) continues to apply to PCCM, DCM, or BCM. In DCM, $I_{DC}=0$. Hence, equation (2.46) reduces to equation (2.45).

In the event of unbalanced load currents (i.e., unequal current across the LED strings), the scalability models continue to hold. The only change is to replace I_{LED} in equations (2.44), (2.45), or (2.46) with $\max(I_{LED})$, where $\max(I_{LED})$ is the *largest* LED current among all strings. In other words, the maximum number of LED strings is constrained by the *largest* LED current. Typically, the input voltage, the output voltage, and the maximum current ripple requirement along with the component values such as the inductor and output capacitors are already defined in the design specifications. So, the only design variable in equations (2.44)–(2.46) is the LED current I_{LED}. In practice, the LED current is often a limiting factor in determining the maximum possible number of outputs in the BCM/DCM/PCCM scalability model. By knowing the maximum LED current required for a particular application, the maximum achievable number of LED strings of a SIMO LED driver can be predicted in advance.

2.5 SUMMARY OF KEY POINTS

1. SIMO DC-DC converters employ only a single inductor in the power stage to supply multiple independent DC outputs. Compared with conventional power management architectures for producing multiple DC output voltages, the SIMO DC-DC converter carries the conspicuous advantages of small form factor, low cost, high efficiency, reduced design complexity, and scalability to multiple outputs.

2. There are four types of power stage topologies of SIMO DC-DC converters, namely buck, boost, inverting buck-boost, and noninverting buck-boost. The SIMO asynchronous and synchronous buck converters require a total of $(N+1)$ and $(N+2)$ power switches, respectively,

where N is the number of outputs. The SIMO boost and inverting buck-boost converters require a total of $(N + 1)$ power switches. The SIMO noninverting buck-boost converter requires a total of $(N + 3)$ power switches.

3. TM hysteretic and PWM control schemes can be used to realize the multiple-energizing switching sequence of a SIMO DC-DC converter. On the other hand, pure hysteretic and hybrid PWM-hysteretic control schemes can be used to realize the single-energizing switching sequence of a SIMO DC-DC converter.

4. The scalability models are formulated for a SIMO LED driver operating in BCM, DCM, and PCCM. Such analytical models enable the prediction of the maximum achievable number of outputs (LED strings) of a SIMO converter under a given set of design parameters.

REFERENCES

1. D. Ma, W.-H. Ki, C.-Y. Tsui, and P. K. T. Mok, "Single-inductor multiple-output switching converters with time-multiplexing control in discontinuous conduction mode", *IEEE J. Solid-State Circuits*, vol. 38, no. 1, pp. 89–100, Jan. 2003.
2. D. Ma, W.-H. Ki, and C.-Y. Tsui, "A pseudo-CCM/DCM SIMO switching converter with freewheel switching", *IEEE J. Solid-State Circuits*, vol. 38, no. 6, pp. 1007–1014, Jun. 2003.
3. D. Kwon and G. A. Rincon-Mora, "Single-inductor-multiple-output switching DC-DC converters", *IEEE Trans. Circuits Syst., II: Exp. Briefs*, vol. 56, no. 8, pp. 614–618, Aug. 2009.
4. W. Xu, Y. Li, X. Gong, Z. Hong, and D. Killat, "A dual-mode single-inductor dual-output switching converter with small ripple", *IEEE Trans. Power Electron.*, vol. 25, no. 3, pp. 614–623, Mar. 2010.
5. X. Jing, P. Mok, and M. Lee, "A wide-load-range constant-charge-auto-hopping control single-inductor- dual-output boost regulator with minimized cross-regulation", *IEEE J. Solid-State Circuits*, vol. 46, no. 10, pp. 2350–2362, Oct. 2011.
6. H. Chen, Y. Zhang, and D. Ma, "A SIMO parallel-string driver IC for dimmable LED backlighting with local bus voltage optimization and single time-shared regulation loop", *IEEE Trans. Power Electron.*, vol. 27, no. 1, pp. 452–462, Jan. 2012.
7. A. Lee, J. Sin, and P. Chan, "Scalability of quasi-hysteretic FSM-based digitally- controlled single-inductor dual-string buck LED driver to multiple string", *IEEE Trans. Power Electron.*, vol. 29, no. 1, pp. 501–513, Jan. 2014.

8. Y. Zhang and D. Ma, "A fast-response hybrid SIMO power converter with adaptive current compensation and minimized cross-regulation", *IEEE J. Solid-State Circuits*, vol. 49, no. 5, pp. 1242–1255, May 2014.

9. W. Sun, C. Han, M. Yang, S. Xu, and S. Lu, "A ripple control dual-mode single-inductor dual-output buck converter with fast transient response", *IEEE Trans. Very Large Scale Integr. Syst.*, vol. 23, no. 1, pp. 107–117, Jan. 2015.

10. C.-W. Chen and A. Fayed, "A low-power dual-frequency SIMO buck converter topology with fully-integrated outputs and fast dynamic operation in 45 nm CMOS", *IEEE J. Solid-State Circuits*, vol. 50, no. 9, pp. 2161–2173, Sep. 2015.

11. Y. Zheng, M. Ho, J. Guo, K.-L. Mak, and K. Leung, "A single-inductor multiple-output auto-buck-boost DC-DC converter with autophase allocation", *IEEE Trans. Power Electron.*, vol. 31, no. 3, pp. 2296–2313, Mar. 2016.

12. K. Modepalli and L. Parsa, "A scalable N-color LED driver using single inductor multiple current output topology", *IEEE Trans. Power Electron.*, vol. 31, no. 5, pp. 3773–3783, May 2016.

13. K. K.-S. Leung and H. S.-H. Chung, "Dynamic hysteresis band control of the buck converter with fast transient response," *IEEE Trans. Circuits Syst., II: Exp. Briefs*, vol. 52, no. 7, pp. 398–402, Jul. 2005.

14. F. Su, W.-H. Ki, and C.-Y. Tsui, "Ultra fast fixed-frequency hysteretic buck converter with maximum charging current control and adaptive for DVS applications", *IEEE J. Solid- State Circuits*, vol. 43, no. 4, pp. 815–822, Apr. 2008.

15. W.-H. Ki, K.-M. Lai, and C. Zhan, "Charge balance analysis and state transition analysis of hysteretic voltage mode switching converters," *IEEE Trans. Circuits Syst., I: Regul. Pap.*, vol. 58, no. 5, pp. 1142–1153, May 2011.

16. P. Li, D. Bhatia, L. Xue, and R. Bashirullah, "A 90–240 MHz hysteretic controlled DC-DC buck converter with digital phase locked loop synchronization", *IEEE J. Solid-State Circuits*, vol. 46, no. 9, pp. 1–12, Sep. 2011.

17. E. Adib and H. Farzanehfard, "Family of soft-switching pwm converter with current sharing in switches", *IEEE Trans. Power Electron.*, vol. 24, no. 4, pp. 979–984, Jan. 2009.

18. F. Luo and D. Ma, "An integrated switching DC-DC converter with dual-mode pulse-train/pwm control", *IEEE Trans. Circuits Syst., II: Exp. Briefs*, vol. 56, no. 2, pp. 152–156, Feb. 2009.

19. C.-Y. Chiang and C-L. Chen, "Zero-voltage-switching control for a PWM buck converter under DCM/CCM boundary", *IEEE Trans. Power Electron.*, vol. 24, no. 9, pp. 2120–2126, Aug. 2009.

20. S. Dietrich, R. Wunderlich, and S. Heinen, "Stability considerations of hysteretic controlled DC-DC converters", *PRIME 2012, 8th Conference on Ph.D. Research in Microelectronics & Electronics*, pp. 79–82, Jun. 2012.

21. S.-C. Koon, Y.-H. Lam, and W.-H. Ki, "Integrated charge-control single-inductor dual-output step-up/step-down converter", *2005 IEEE International Symposium on Circuits and Systems*, pp. 3071–3074, Jul. 2005.

22. M.-H. Huang and K.-H. Chen, "Single-inductor multiple-output (SIMO) DC-DC converters with high light-load efficiency and minimized cross-regulation for portable devices", *IEEE J. Solid-State Circuits*, vol. 44, no. 4, pp. 1099–1111, Apr. 2009.
23. Application Note: AN-1656, "Design challenges of switching LED divers," Texas Instruments Incorporated (2013). [Online] Available: https://www.ti.com/lit/an/snva253a/snva253a.pdf?ts=1619334368514&ref_url=https%253A%252F%252Fwww.google.com%252F.
24. R. W. Erickson and D. Maksimovic, *Fundamentals of Power Electronics*, 2nd ed. New York, NY, USA: Springer, 2001.

PROBLEMS

2.1 An ideal SIDO DC-DC buck converter operating in DCM shown in Figure 2.19 is designed in accordance with the following specifications:

Input voltage	$V_{in} = 24\,V$
Main inductor	$L = 5\,\mu H$
Output capacitor	$C_{o1} = C_{o2} = 33\,\mu F$
Load resistor	$R_{L1} = R_{L2} = 10\,\Omega$
Switching frequency of main switch Q_1	$f_{sw} = 200\,kHz$
On-time duty ratio of main switch Q_1	$D_{on} = 20\%$

All losses may be ignored. S_{main} is the gate drive signal of Q_1. Likewise, S_{o1} and S_{o2} are the gate drive signals of Q_{o1} and Q_{o2}, respectively.

a. Sketch the theoretical waveforms of S_{main}, S_{o1}, and S_{o2} alongside the inductor current I_L in steady-state condition.

b. Calculate the output voltage of the first output channel V_{o1} of the SIDO converter.

c. Compute the average value of the inductor current.

d. Determine the peak value of the inductor current.

2.2

a. Determine the maximum possible output voltage of the first output channel of the SIDO converter shown in Figure 2.19 without cross-regulation based on the design specifications given in Problem 2.1.

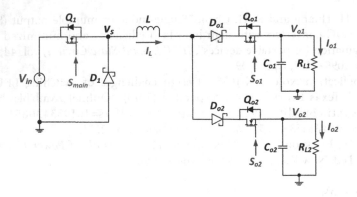

FIGURE 2.19 Ideal SIDO DC-DC buck converter.

b. Based on the result in part (a), calculate the maximum average value of the inductor current.

c. Based on the results in part (a) and (b), compute the maximum peak value of the inductor current.

2.3

a. Create a simulation model for the SIDO DC-DC buck converter shown in Figure 2.19. Use the design specifications given in Problem 2.1. Use simulation software (e.g., Pspice, PSIM, Simulink, etc.) to plot the steady-state waveforms of the gate drive signal of the main switch (S_{main}), the gate drive signals of the output switches (S_{o1}, S_{o2}), and the inductor current (I_L), as well as the two output voltages (V_{o1}, V_{o2}).

b. Use the simulation model in part (a) to obtain the following quantities:

 i. the value of the first output voltage V_{o1} of the SIDO converter,

 ii. the average value of the inductor current, and

 iii. the peak value of the inductor current.

Compare the above simulated values with the corresponding calculated values in Problem 2.1 (b)–(d).

c. Use the simulation model in part (a) to determine the maximum on-time duty ratio of the main switch of the SIDO converter without cross-regulation. Plot the steady-state waveforms of the gate drive signal of the main switch (S_{main}), the gate drive signals of the output switches (S_{o1}, S_{o2}), and the inductor current (I_L), as well as the two output voltages (V_{o1}, V_{o2}).

d. From the simulation results in part (c), obtain the following quantities:

    i. the maximum output voltage of the first output channel of the SIDO converter,

    ii. the maximum average value of the inductor current, and

    iii. the maximum peak value of the inductor current.

Compare the above simulated values with the corresponding calculated values in Problem 2.2 (a)–(c).

2.4   Consider an ideal SIDO DC-DC boost converter operating in PCCM with the following specifications.

Input voltage	$V_{in} = 3\,V$
Main inductor	$L = 3\,\mu H$
Output capacitor	$C_{o1} = C_{o2} = 33\,\mu F$
Load resistor	$R_{L1} = R_{L2} = 50\,\Omega$
Switching frequency of the main switch	$f_{sw} = 200\,kHz$
On-time duty ratio of the main switch	$D_{on} = 30\%$
DC offset of the inductor current	$I_{L,\,dc} = 0.76\,A$

Note: Under balanced load condition, the two DC output voltages of the SIDO converter are assumed to be identical.

a. Sketch the circuit diagram of this ideal SIDO DC-DC boost converter in PCCM.

b. Derive a general expression for the peak value of the inductor current of the SIDO DC-DC boost converter operating in PCCM. Based on the above specifications, find the peak value of the inductor current.

   c. Derive a general expression of the output voltage of the SIDO DC-DC boost converter.

   d. Based on the above specifications and the expression of the output voltage derived in part (c), determine the output voltage of the SIDO DC-DC boost converter.

   e. Find the average value of the inductor current.

2.5

   a. Create a simulation model of the SIDO DC-DC boost converter operating in PCCM based on the design specifications in Problem 2.4. Use a simulation software (e.g., Pspice, PSIM, Simulink, etc.) to plot the steady-state waveforms of the gate drive signal of the main switch, the gate drive signals of the two output switches, the gate drive signal of the freewheeling switch, the inductor current (I_L), and the two output voltages.

   b. From the simulation results in part (a), obtain the following quantities:

      i. the peak value of the inductor current,

     ii. the average value of the inductor current, and

    iii. the average value of the output voltage.

Verify that the simulated and calculated values of each of the above parameters are in close agreement with each other.

2.6   Consider an ideal SIMO DC-DC buck converter operating in BCM with the following specifications:

Input voltage	$V_{in} = 12\,V$
Main inductor	$L = 2.4\,\mu H$
Output capacitor per output	$C_o = 10\,\mu F$
Load resistor per output	$R_L = 10\,\Omega$
Switching frequency of the main switch	$f_{sw} = 300\,kHz$

You can safely assume that the SIMO DC-DC buck converter operates under balanced load condition in which the DC output voltages of the SIMO converter are the same.

a. Suppose the maximum output voltage ripple is 4% of the nominal value of the output voltage. Determine the theoretical maximum achievable number of outputs of the SIMO DC-DC buck converter in BCM.

b. Create a simulation model for the SIMO DC-DC buck converter with N outputs, where N is the theoretical maximum possible number of outputs obtained in part (a).

Use simulation to show that the maximum output voltage ripple is less than 4% of the nominal value of the output voltage.

2.7   Consider a SIDO DC-DC buck converter with the same specifications given in Problem 2.6, except that the value of the output capacitor is increased from 10 to 33 μF.

a. Discuss the ramification of increasing the value of the output capacitor. Perform simulation of this SIMO converter. Are the simulation results consistent with your expectation? Explain your answer.

b. Assume that the maximum output voltage ripple is 3% of the nominal value of the output voltage. Determine the theoretical maximum achievable number of outputs of a SIMO DC-DC buck converter operating in BCM when the value of the output capacitor is 33 μF.

c. Use simulation to verify your result in part (b). You may construct a simulation model of a SIMO DC-DC buck converter with N outputs, where N is the theoretical maximum number of outputs obtained in part (b).

CHAPTER 3

Single-Inductor Multiple-Output AC-DC Converters

3.1 INTRODUCTION

Direct current (DC) and alternating current (AC) are two vastly different approaches to the transmission and distribution of electric power. DC current works by supplying a constant electric voltage to the DC-powered devices or equipment, which draw a constant DC electric current. Battery is a common source of DC, and most portable electronic devices require DC current to operate. Alternatively, AC current keeps switching directions periodically. It flows forwards and backwards due to an oscillating voltage with a line frequency of 50 or 60 Hz. In fact, nearly all commercial electricity has been produced and transmitted via AC over the past century. This is mainly because of the fact that the principle of magnetic induction only works for AC. What really makes AC superior in terms of power transmission is that based on magnetic induction, it is fairly easy and inexpensive to step up (or down) the voltage by means of an iron-core transformer. Besides, since the power loss due to cable resistance along a transmission line increases as the square of the current increases (or inversely proportional to the square of the voltage), it is highly desirable to employ high-voltage power transmission in order to significantly mitigate power losses. AC-DC conversions are therefore needed to interface

DOI: 10.1201/9781003239833-3

the downstream DC loads to the existing AC power grids. For example, if AC is the type of power delivered to your home and DC is the type of power needed to switch on your LED lamp or charge your smartphone, you will need an AC-DC power supply to convert the mains voltage from the power grid to a constant and stable DC voltage or current as required by the end device or equipment. Hence, it is beneficial to design compact, efficient, and cost-effective AC-DC power supplies.

In this chapter, an offline (nonisolated) single-stage single-inductor multiple-output (SIMO) AC-DC converter topology is introduced [1,2]. It can simultaneously supply multiple DC loads (e.g. LEDs) from the AC mains. A key application of this converter is that it can act as an offline AC-DC LED driver for supplying multiple independent LED strings in LCD backlighting and general lighting or color-mixing applications. Such transformerless LED driver topology carries conspicuous advantages such as high compactness, light weight, and low cost, which are well suited for space-constrained retrofit lighting applications. Without loss of generality, each output of the SIMO AC-DC converter is assumed to be connected to an LED load for illustration purpose. Nonetheless, it can also act as an AC-DC adapter for supplying multiple low-power rating devices or equipment such as cell phones, laptops, tablets, wearables, DVD players, battery-powered vacuum cleaner, and Internet-of-Things gadgets.

Until recently, the SIMO architecture was used almost exclusively for DC-DC power conversion [3–12]. Due to its simple and scalable circuit architecture, a number of SIMO DC-DC high-brightness LED drivers have quickly emerged, which are originally geared toward battery-operated portable electronic devices [6,7,10]. They provide independent current regulation to each individual LED string from a single DC power source. But the fact of the matter is that the majority of the lighting fixtures or luminaires for general illumination purpose ultimately derive their power from the AC mains. Thus, if a SIMO DC-DC LED driver is used in such cases, it needs to be preceded by an AC-DC converter, which transforms the AC mains voltage (e.g., 110 V/60 Hz or 220 V/50 Hz) to an intermediate DC-link voltage. This leads to the formulation of a two-stage AC-DC SIMO topology [13,14]. Figure 3.1 shows the system architecture of this AC-DC SIMO LED driver, which comprises the first-stage AC-DC converter in series with the second-stage SIMO DC-DC converter.

The AC-DC front stage can be as simple as a diode bridge rectifier in series with a large DC-link capacitor [13]. Aluminum electrolytic

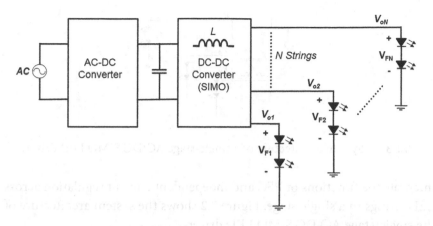

FIGURE 3.1   System architecture of a two-stage AC/DC SIMO LED driver.

capacitors (E-caps) are a common choice for DC-link capacitor because they are inexpensive and have high capacitance values, albeit limited lifetime. Unfortunately, this nonisolated AC-DC converter produces an unregulated DC bus voltage without power factor (PF) correction (PFC). As a result, such kind of AC-DC converter can only be used for low-power LED lighting applications, of which the PF requirement is less stringent [15,16]. Subsequently, another two-stage AC/DC SIMO LED driver is reported in [14]. It consists of a first-stage PFC boost converter and a second-stage SIMO buck converter. Even though this LED driver can produce a well-regulated DC bus voltage while achieving a high PF of 0.99, the fact that the SIMO converter operates in continuous conduction mode (CCM) means that it is susceptible to undesirable cross-regulation issues. In other words, a change in the current of one LED channel induces an unwanted change in the current of another LED channel. Hence, independent current regulation across the LED channels is *not* feasible. Only current balancing function is performed. Besides, two separate sets of controllers are needed for the two-stage LED driver, namely, one for the PFC boost AC-DC stage and the other for the SIMO DC-DC stage, which further complicates the system design. This two-stage circuit structure also requires the use of a DC-link capacitor. If the DC-link voltage is high, it is rather difficult to select an appropriate capacitor that has a long lifetime and low cost. The use of short-lifetime E-cap undermines the overall robustness and reliability of the LED driver. The aforementioned issues have prompted researchers to explore a new single-stage SIMO topology for driving multiple LED strings. The main challenge is to properly

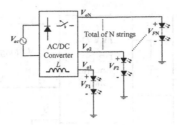

FIGURE 3.2   System architecture of a single-stage AC/DC SIMO LED driver.

integrate the functions of PFC and independent current regulation across LED strings in a single stage. Figure 3.2 shows the system architecture of the single-stage AC-DC SIMO LED driver.

3.2 CIRCUIT TOPOLOGY

Unlike its two-stage counterpart, this one-stage SIMO LED driver can drive multiple independent LED strings directly from the AC mains supply. The functions of PFC and independent regulation of string currents are simultaneously achieved. This is made possible through the proper component integration of a PFC stage and a DC-DC SIMO converter. Figure 3.3 shows the formulation of a single-stage SIMO AC-DC buck-type LED driver. For the sake of discussion, suppose we have a buck PFC converter operating in discontinuous conduction mode (DCM) [see Figure 3.3a] and a buck-type SIMO DC-DC converter [see Figure 3.3b]. Careful observation reveals that the two converters do share some similarities. Most notably, they share a common circuit configuration that includes the main power switch (S_a, S'_a), the freewheeling diode (D_a, D'_a), and the inductor (L, L'). They also share a common ground. Hence, the two converters can readily be combined together to form a novel single-stage SIMO AC-DC topology, as shown in Figure 3.3c.

By employing the time-multiplexing switching scheme, each LED channel of the single-stage SIMO LED driver operating in DCM is fully decoupled in time. Thus, the SIMO LED driver behaves just like a single-input single-output (SISO) LED driver at any instance in time. Since a buck converter in DCM is naturally an emulated resistor at low frequencies [17], the averaged input current of the LED driver over a switching period is inherently proportional to the line voltage. Consequently, the original DC-DC SIMO LED driver can easily be turned into a single-stage AC-DC SIMO converter with built-in PFC function by adding a front-end

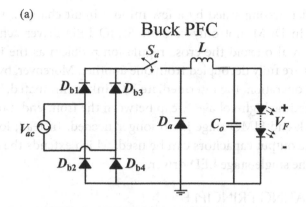

(a)

Buck PFC

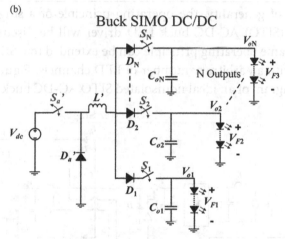

(b)

Buck SIMO DC/DC

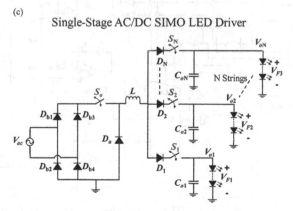

(c)

Single-Stage AC/DC SIMO LED Driver

FIGURE 3.3 Formulation of a single-stage SIMO AC-DC buck-type LED driver. (a) Buck PFC converter in DCM; (b) buck SIMO DC-DC LED driver; and (c) single-stage SIMO AC-DC LED driver.

diode rectifier accompanied by a few minor circuit changes. By operating strictly in DCM, not only can the SIMO LED driver achieve high PF, but it can also avoid the cross-regulation problem as the individual LED strings are fully decoupled from one another. Moreover, by enabling single-stage operation, the intermediate DC link is eliminated. Therefore, the short-lifetime high-voltage E-cap between the front-end diode bridge and the back-end SIMO stage is no longer needed. Instead, low-voltage long-lifetime output capacitors can be used, which extends the operating lifetime of the single-stage LED driver.

3.3 OPERATING PRINCIPLE

Without loss of generality, the operating principle of a single-inductor three-output (SITO) AC-DC buck LED driver will be discussed in this section. The same operating principle can be extended to a SIMO AC-DC LED driver with an arbitrary number of LED channels. Figure 3.4 shows the circuit diagram of an ideal nonisolated SITO AC-DC buck LED driver

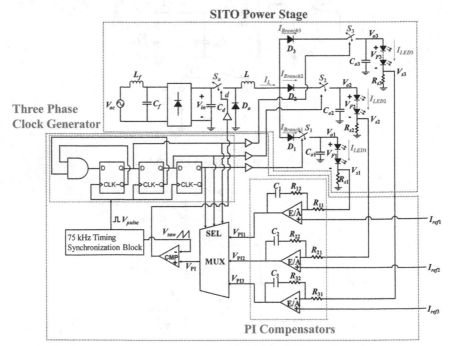

FIGURE 3.4 Circuit diagram of an ideal SITO AC-DC buck LED driver with closed-loop control.

with closed-loop control. The SITO LED driver comprises two major functional blocks, i.e., the power stage and the time-multiplexed feedback controller. The operations of these two blocks will be elucidated in the following subsections.

Before delving into the operation of the power stage, it is absolutely important to have a clear understanding of the definitions of *balanced load* and *unbalanced load* and be able to distinguish the difference between them. The words "balanced load" and "unbalanced load" appear quite frequently throughout the book and can sometimes be confusing to readers. In general terms, balanced load refers to the specific condition in which the outputs of the SIMO converter have the *same* current and voltage value. Simply put, all outputs of the SIMO converter have the *same* power rating. For illustration purpose, let us use the SITO LED driver as an example. The SITO LED driver supplies three independent LED strings connected in parallel. To satisfy the balanced load condition, the three strings should have the same output power. This can be accomplished by using the same type and same number of LEDs (e.g., six blue LEDs from the same bin) per string and ensuring that all three strings carry the same current. On the other hand, unbalanced load refers to the general condition in which the outputs of the SIMO converter have *different* current and/or voltage values. This means that the outputs have different power ratings. Again, we use the SITO LED driver as an example. One way to satisfy the unbalanced load condition is to connect the same type and same number of LEDs and provide a *unique* average current per string. Alternatively, different types of LEDs (e.g., RGB LEDs) with distinct I-V characteristics can be used across the three strings. Even if the same current value flows across the three strings, their voltages will still be different. In either case, the LED channels will have different output power levels.

3.3.1 Power Stage

The power stage of this SITO LED driver with three parallel LED strings requires a total of *four* power switches, including one main switch (S_a) and three output switches (S_1, S_2, S_3). A simple L_f–C_f input EMI filter is added at the power input of the LED driver to attenuate the switching harmonics that are present at the input current waveform. By employing a full-wave rectifier (e.g., single-phase H-bridge rectifier), the AC voltage at the output of the EMI filter is converted into a pulsating DC voltage that oscillates at twice the line frequency. Instead of using a large DC-link

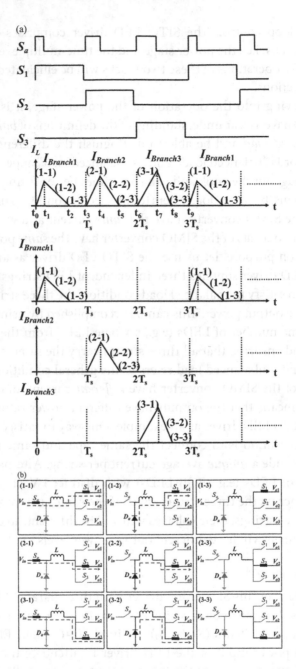

FIGURE 3.5 (a) Theoretical waveforms of the gate drive signals of the main switch (S_a) and the three output switches (S_1–S_3) along with the inductor current I_L, and the three branch currents ($I_{branch1}$–$I_{branch3}$); (b) switching sequence of the SITO AC-DC LED driver.

capacitor to absorb the double-line frequency ripple power, a small high-frequency filter capacitor C_d is added at the output of the bridge rectifier to reduce the switching ripples in the DC voltage, where the DC voltage is the same as the input voltage (V_{in}) of the backend SITO buck converter. The rectified current then flows into the power stage of the SITO converter. Note that only a single inductor L is used in the power stage. D_a is the freewheeling diode. D_i is the branch diode associated with the ith LED string, which eliminates undesirable reverse current flow across the output switch, where $i = 1, 2,$ and 3. Suppose each ideal switch is realized using silicon-based N-channel power MOSFET. Although the MOSFET is switched OFF, a reverse current can still flow across the MOSFET via its intrinsic body diode when the voltage at the source terminal is higher than the voltage at the drain terminal. Let us consider the following scenario. When the first output is enabled and the second output voltage becomes *larger* than the first output voltage (i.e. $V_{o2} > V_{o1}$) in unbalanced load condition, an unwanted reverse current will flow from the second output to the first output via the intrinsic body diode of the second output MOSFET. The two outputs are virtually shorted together. Hence, the branch diodes are needed to prevent cross-interference among the output channels. C_{oi} and R_{si} are the output capacitor and sensing resistor of the ith LED string, respectively. Since this is a buck converter whose output voltage is always lower than the RMS value of the AC input voltage, a low-voltage high-value capacitor can be used as the output capacitor to suppress the double-line frequency ripples and high-frequency ripples and thus incurs lower cost, as compared to using a high-voltage high-value capacitor in the DC-link. The AC mains voltage is denoted by V_{ac}. The rectified voltage after the diode bridge is represented by V_{in}, and the first, second, and third output voltages are V_{o1}, V_{o2}, and V_{o3}, respectively. I_L is the current across the main inductor L. $I_{branch1}$, $I_{branch2}$, and $I_{branch3}$ are the individual branch currents that flow across the first, second, and third output switches, respectively. Figure 3.5a depicts the theoretical waveforms of the gate drive signals of S_a and S_1–S_3 as well as I_L and $I_{branch1}$–$I_{branch3}$. Note that T_s represents the switching period of the main switch S_a. The above notations will be used throughout the rest of the chapter.

Figure 3.5b shows the switching sequence of the SITO AC-DC LED driver. Basically, the energy stored in the inductor L is transferred to each of the three outputs (or LED channels) successively in a round-robin time-multiplexed manner. Only a particular output will be enabled per switching cycle. Such switching scheme is also known as multiple energizing phases. For three consecutive switching cycles, e.g., from time $t = 0$ to $t = 3Ts$, there are a total of nine distinct switching states represented by (x, y), where x is the output number ($x = 1, 2, 3$), and y is the mode of operation ($y = 1, 2, 3$).

In DCM, there are *three* modes of operation, each of which is described in more detail below. Without loss of generality, we will consider the first switching cycle pertaining to the first output for illustration purpose.

Mode 1 (t_0–t_1): The main switch (S_a) and the first output switch (S_1) are switched ON, while the freewheeling diode D_a is OFF. The second and third output switches (S_2, S_3) are OFF since only the first output is enabled. The inductor L is charged up, and the inductor current I_L *increases* linearly at a rate of $(V_{in}-V_{o1})/L$. This is often referred to as the first sub-interval of DCM (or the charging phase). This switching state is uniquely represented as (1–1). Likewise, (2–1) represents the second output and mode 1, while (3–1) represents the third output and mode 1. The time duration of Mode 1 is ultimately determined by the on-time duty ratio of S_a for the first output. At $t = t_1$, I_L reaches its maximum (peak) value.

Mode 2 (t_1–t_2): S_a is switched OFF, S_1 remains ON, and D_a is ON. As the inductor discharges, I_L *decreases* linearly at a rate of V_{o1}/L. At $t = t_2$, I_L drops to zero value as the inductor is fully discharged. This marks the end of Mode 2 and the beginning of Mode 3. This is also referred to as the second sub-interval of DCM (or the discharging phase). This switching state is uniquely represented as (1–2). Similarly, (2–2) represents the second output and mode 2, while (3–2) represents the third output and mode 2.

Mode 3 (t_2–t_3): Both S_a and D_a are OFF. S_1 remains ON. Alternatively, to reduce switching loss, S_1 can be switched OFF early with zero-current switching. This is commonly known as the third sub-interval of DCM (or the idle phase). Ideally, the inductor current I_L stays at zero value during this idle interval. This switching state is distinctively represented as (1–3). Similarly, (2–3) represents the second output and mode 3, while (3–3) represents the third output and mode 3.

The same switching process is then repeated in the subsequent two switching cycles for the second and third output, respectively, during which S_1 is OFF, while S_2 and S_3 are switched ON sequentially. Hence, the energy is distributed from the inductor to the three outputs in a time-interleaved manner. In general, the same switching sequence can easily be extended to N outputs in a SIMO AC-DC converter, where N is an integer.

The output switch corresponding to a particular output is switched ON only during one of the N switching cycles, and it is switched OFF during the remaining $(N-1)$ switching cycles.

3.3.2 Time-Multiplexed Controller

The AC/DC SITO buck LED driver is regulated by an analog-based time-multiplexed controller as shown in Figure 3.4. Based on the operating principle discussed in the previous section, the rising edge of the gate drive signal of S_a should be synchronized with that of the output switches S_1–S_3. The synchronization is realized by the clock generation block, which will be discussed in Section 3.3.3. The average current of each LED string is controlled by the corresponding control loop. The current-sense voltage (V_{si}), which is directly proportional to the LED current, is subtracted from a fixed reference value (I_{refi}) to generate an error signal (V_{EAi}), where the subscript i denotes the LED channel (i.e., $i = 1$, 2, or 3). This error signal is compensated by a conventional PI controller and then modulated by a PWM modulator to yield the on-time duty ratio of the main switch (S_a). A three-phase clock generator is used to derive three nonoverlapping clock signals running at 25 kHz, which is used to select one of the three outputs at any time instance. Figure 3.6 shows the time-multiplexed PWM control scheme using trailing-edge modulation. It assumes that three distinct-colored LEDs (e.g. red, geen, and blue LEDs) with different I-V characteristics are used across the three LED strings. Basically, the figure shows that the main switch (S_a) is turned OFF when

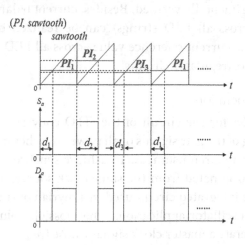

FIGURE 3.6   Time-multiplexed control scheme using trailing-edge modulation.

the modulating ramp (in this case, the sawtooth waveform) reaches the level of the PI output. In this way, the on-time duty ratio of S_a can be determined. It should be noted that for unidentical LED loads and/or different current references, the steady-state values of the three PI outputs will *not* be the same, and thus, the on-time duty ratios across the LED strings will also be different. To reduce the number of PWM modulators and logic elements, the outputs of the PI compensators are time-multiplexed in such a way that a common PWM modulator (rather than three parallel PWM modulators) can be used. In other words, the PWM modulator is shared among the three PI compensators. The use of a single PWM modulator also enables the sharing of subsequent digital logic among the three outputs. Obviously, this carries the advantages of reduced component count, high compactness, lower build-of-material (BOM) cost, and increased efficiency, which is especially appealing for space-constrained and cost-sensitive applications such as MR-16, GU-10 retrofits, and other similar solid-state lighting products.

The use of time-interleaving control in conjunction with multiple energizing phases means that each LED string is independently driven and decoupled from the other strings with minimal cross-interference. The current in each individual LED string can be regulated separately by assigning a unique current reference for each string. It is envisaged that under different loading conditions and/or current reference commands, the inductor current (I_L) will exhibit different slopes and peak values across the LED channels. In fact, such phenomenon is illustrated in Figure 3.5a and will be experimentally verified. Besides, current balancing (i.e., same current value across all LED strings) can be realized conveniently by applying the same current reference value across all LED strings without the addition of postregulator circuits.

3.3.3 Clock Generation

Note that in order for the circuit of the SITO converter to operate correctly, the timing of the gate signals of all power switches must be well synchronized under all circumstances. To achieve synchronization, the gate signals must be originated from the same clock source. The 555 timer is indeed a popular integrated circuit used in a myriad of timer, delay, pulse generation, and oscillator applications. One possible solution is to use a 555 timer to generate a master clock signal whose frequency is identical to the switching frequency of the SITO converter. In addition to the master clock, the 555 timer can also be configured in such a way that a sawtooth waveform running at the same frequency as the master clock can be generated as well. Recall that this sawtooth waveform is required by the PWM modulator in the controller. Figure 3.7 shows the configuration of the 555 timer in monostable state.

The bias voltage of the bipolar junction transistor (T), as shown in Figure 3.7, is set by a simple voltage divider formed by two resistors in series, namely, R_{T2} and R_{T3}. R_{T1} is used to limit the current flowing through the BJT for charging the capacitor C_{T1}. Hence, the voltage across C_{T1} is given by

$$V_{CT1} = \frac{Q}{C_{T1}} = \frac{I_T}{C_{T1}} t \qquad (3.1)$$

where Q is the charge stored in C_{T1}, and I_T is the current across the BJT.

Under this external circuit configuration, the 555 timer produces a smooth and almost linear ramp signal (V_{saw}) at pin 6, as shown in Figure 3.7. Pin 3 of the 555 timer will generate a negative-going pulse every time the capacitor C_{TI} discharges. Hence by inverting this pulse, a master clock signal (V_{pulse}), which is synchronized with V_{saw}, can be generated. V_{pulse} serves as the clock signal of the three-phase clock generator to generate the non-overlapping gate signals of the three output switches as well as the set signal for triggering the leading edge of the gate signal of the main switch. On the other hand, V_{saw}, the sawtooth signal, is fed to the PWM modulator to determine the trailing edge of the gate signal of the main switch.

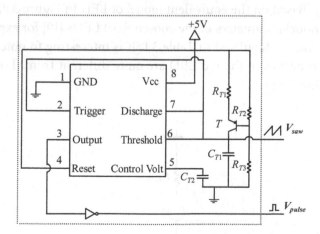

FIGURE 3.7 555 timer in monostable state generates a linear ramp V_{saw} and a clock signal V_{pulse}.

3.4 DESIGN CONSIDERATION

3.4.1 Inductor Design

To simplify the controller design for PFC and also minimize the size of the inductor, the SITO converter needs to operate in DCM. Another design consideration is to limit the current ripple across the inductor in order to reduce the current stress of the power switches. Consequently, the value of the inductor should neither be too large nor too small. Therefore, in this subsection, the lower bound and upper bound of the inductor value will be analytically derived.

Let us digress for a moment and discuss about the LED model. It is because the LED load also plays a significant role in the inductor design. Figure 3.8 shows the equivalent circuit model of an LED. It is also referred to as an approximate linear model which predicts the forward voltage as a function of the forward current of LED [18]. This model is reasonably accurate only within the linear region of the actual I-V curve of LED. Basically, it contains a series connection of an ideal diode (D_{LED}), a resistor (R_{LED}), and a threshold voltage (V_{th}). D_{LED} is used to model the unidirectional current flow of an LED. R_{LED} is the small-signal resistance of an LED, which is equal to the reciprocal of the slope of the *linear* portion of the I-V curve, and V_{th} is the "turn-on" voltage of LED.

Because the SITO AC-DC converter can simultaneously supply three individual LED strings, it can act as a red-green-blue (RGB) LED driver for color-mixing applications. For example, the first, second, and third output of this LED driver can be connected to red, green, and blue LEDs, respectively. Based on the equivalent model of LED in Figure 3.8, the values of the model parameters of the chosen RGB LEDs [19] for experimental verification are tabulated in Table 3.1. It is interesting to note that the electrical properties of the red LED are quite different from those of the green (or blue) LED.

FIGURE 3.8   Equivalent circuit model of an LED.

TABLE 3.1   Model Parameters of the Chosen RGB LEDS

Model Parameter	Red LED	Green LED	Blue LED
Equivalent resistance, R_{LED} (Ω)	4	6	6
Rated current, I_{LED} (mA)	350	350	350
Threshold voltage V_{th} (V)	0.7	0.8	0.85
Forward voltage, V_F (V)	2.1	2.9	2.95
Rated power, P_{LED} (W)	0.735	1.015	1.0325

Now, let us consider a general case of a single-inductor multiple-output (SIMO) AC-DC converter which supplies N number of outputs. Further, we assume that the LED loads are nonidentical (unbalanced) across the outputs. Due to the unbalanced load condition, the design constraint of the inductor (L) from one output will be different than that from another LED load. To simplify the analysis, we can decompose the SIMO converter into N number of SISO converters because the outputs of a SIMO converter operating in DCM are independent of each other. Then, each SISO converter with its corresponding LED load can be analyzed relatively easily. To start with, let us consider an SISO DC-DC buck converter operating in CCM in steady-state condition. Figure 3.9 shows the ideal waveform of the voltage across the inductor of a SISO DC-DC buck converter in CCM. By invoking the volt-second balance of an inductor, the following general equation can be written for N SISO buck converters.

$$(V_{in} - V_{oi})d_iT_s - V_{oi}(1 - d_i)T_s = 0 \qquad (3.2)$$

where V_{in} is the input voltage, V_{oi} is the output voltage, and d_i is the duty ratio of the ith SISO buck converter, where $1 \leq i \leq N$.

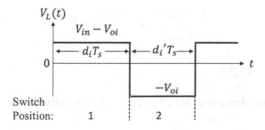

FIGURE 3.9   Ideal waveform of the voltage across the inductor of a DC-DC buck converter in CCM.

The output voltage of the ith buck converter can be written as

$$V_{oi} = d_i V_{in} \tag{3.3}$$

In the first switching cycle, the rate of increase of the inductor current I_{Li} is given by

$$\frac{dI_{Li}}{dt} = \frac{V_{Li}(t)}{L_i} = \frac{V_{in} - V_{oi}}{L_i} \tag{3.4}$$

where L_i is the inductance of the ith buck converter.

The peak-to-average current ripple is defined as

$$\Delta I_{Li,\,pa} = \left(\frac{V_{in} - V_{oi}}{2L_i} \right) d_i T_s. \tag{3.5}$$

Under steady-state condition, the DC component of the capacitor current is ideally zero. Therefore, the DC component of the inductor current can be expressed as

$$I_{Li} = I_{LEDi} = \frac{V_{oi} - m V_{thi}}{m R_{LEDi}} \tag{3.6}$$

where m is the number of LEDs at the output of the ith buck converter. I_{LEDi}, V_{thi}, and R_{LEDi} are the rated current, the threshold voltage, and the small-signal resistance of a single LED, respectively.

In particular, when the system operates in DCM, $I_{Li} < \Delta I_{Li}$, where ΔI_{Li} represents the maximum inductor current ripple when the buck converter operates in boundary conduction mode (BCM), also known as critical conduction mode. That is,

$$\frac{V_{oi} - m V_{thi}}{m R_{LEDi}} < \frac{V_{in} - V_{oi}}{2L_i} d_i T_s \tag{3.7}$$

where $d_i = \dfrac{V_{oi}}{V_{in}}$ in BCM.

Hence, the maximum value of L_i (L_{i_max}) is obtained as

$$L_{i\_max} = \frac{(V_{in} - V_{oi}) m R_{LEDi}}{2(V_{oi} - m V_{thi})} d_i T_s = \frac{(V_{in} - V_{oi}) m R_{LEDi}}{2(V_{oi} - m V_{thi})} \frac{V_{oi}}{V_{in}} T_s \tag{3.8}$$

By taking into account *all* SISO buck converters, the maximum allowable value of the inductor (*L*) of the SIMO buck converter is therefore given by

$$L \leq min\{L_{i\_max}\}, \quad i = 1, 2, 3, ..., N \tag{3.9}$$

Now, consider the minimum allowable value of *L*. It can be obtained by examining the maximum allowable ripple of the inductor current (ΔI_{L_max}). Mathematically, we can write

$$\Delta I_{Li} = \left(\frac{V_{in} - V_{oi}}{L_i} \right) d_i T_s \leq \Delta I_{L\_max} \tag{3.10}$$

In DCM operation, the output (LED) power can be expressed as

$$P_{LEDi} = \frac{V_{rms}^2}{R_{e\_i}(d_i)} \tag{3.11}$$

where V_{rms} is the RMS value of V_{in} (Note: V_{in} is a pulsating DC voltage in the SIMO AC-DC converter), and $R_{e\_i}(d_i)$ is the equivalent resistance as seen by the input port of the ith buck converter, which is given by [17]

$$R_{e\_i}(d_i) = \frac{2L_i}{d_i^2 T_s} \tag{3.12}$$

Notice that P_{LEDi} is the power consumed by the LED load at the output of the ith buck converter, which can be expressed as

$$P_{LEDi} = m V_{Fi} I_{LEDi} \tag{3.13}$$

where V_{Fi} is the forward voltage of LED of the ith buck converter.

From equations (3.11) and (3.12), the duty cycle d_i can be derived as follows.

$$d_i = \sqrt{\frac{2L_i}{R_{e\_i}(d_i)T_s}} = \sqrt{\frac{2L_i P_{LEDi}}{V_{rms}^2 T_s}} \tag{3.14}$$

By substituting equation (3.14) into (3.10), the minimum value of L_i can be expressed as

$$L_{i\_min} = \left(\frac{V_{in} - V_{oi}}{\Delta I_{L\_max}} \right)^2 \frac{2 P_{LEDi} T_s}{V_{rms}^2} \tag{3.15}$$

By taking into account *all* SISO buck converters, the minimum allowable value of the inductor (L) of the SIMO buck converter is therefore given by

$$L \geq max\{L_{i\_min}\}, \quad i = 1,2,3,...,N \tag{3.16}$$

Based on equations (3.9) and (3.16), the range of acceptable values of L can be determined.

3.4.2 Output Capacitor Design

Besides the sizing of the inductor, it is also important to choose an optimal value of the output capacitor of the SIMO AC-DC converter. Similar to the design methodology for the inductor, the SIMO converter can first be decomposed into N number of SISO converters due to independent operation of each output of the SIMO converter. Then, the design of the output capacitor can be performed on each of the SISO converters. In fact, the design approach of the output capacitor resembles that for a DC-link capacitor in conventional AC-DC rectifying systems since both have to perform the same functions of AC energy storage of the pulsating power due to double-line frequency and the mitigation of the high-frequency ripples. This is different from that of the DC-DC SIMO converter in which the output capacitor is used to suppress only high-frequency (switching) ripples.

If $\Delta V_o = k V_{oi}$, where V_{oi} is the average value of the output voltage of the ith SISO buck converter, ΔV_o is the peak-to-peak value of the output voltage ripple, and k is the ripple factor that defines the allowable peak-to-peak voltage ripple. From [17], the lower limit for C_{oi} is given by

$$C_{oi} \geq \frac{P_{LEDi}}{k V_{oi}^2} \times \frac{T_{line}}{2\pi} \tag{3.17}$$

where T_{line} is the line period (e.g., $T_{line} = 1/60\,Hz$, assuming the line frequency is $60\,Hz$).

If we were to choose the same value for the output capacitors across all outputs of the SIMO AC-DC converter, the optimal value would be determined by the *largest* value among all output capacitors. Mathematically, we can write

$$C_o \geq max\{C_{oi}\}, \quad i = 1,2,3,...,N \tag{3.18}$$

3.4.3 Controller Design

Due to the inherent nature of the time-multiplexing control scheme, only one feedback loop of the SIMO converter is active at any point in time, albeit multiple feedback loops, one per output, exist in the SIMO system. By taking advantage of the fact that the PI compensator of one output is fully decoupled from that of the other outputs, we can effectively decompose the SIMO buck converter with a single time-multiplexed controller into multiple independent SISO buck converters with their respective controllers. Consequently, the small-signal analysis can be simplified by considering the open-loop transfer function of the SISO buck converter with a single feedback loop. Figure 3.10 shows the small-signal block diagram of the ith buck converter, where $1 \leq i \leq N$, and N is the total number of SISO buck converters. The ith buck converter corresponds to the ith output or, more precisely, the ith LED string of the SIMO LED driver.

The power plant in the closed-loop system is basically a buck converter operating in DCM. As a simple approximation, the small-signal control-to-output transfer function of the DCM buck converter pertaining to the ith string, denoted by $G_{buck\_i}(s)$, can be analytically derived by letting the value of the inductor (L) tend to zero [17]. Hence, the expression of $G_{buck\_i}(s)$ can be obtained as follows.

$$G_{buck\_i}(s) = \frac{\hat{i}_{oi}}{\hat{d}_i}\bigg|_{\hat{v}_g = 0} = \frac{\dfrac{2V_{oi}}{d_i} \times \dfrac{1-M_i}{2-M_i} \times \dfrac{1}{mR_{LEDi}}}{1 + \dfrac{s}{\dfrac{2-M_i}{(1-M_i)mR_{LEDi}C_{oi}}}} \qquad (3.19)$$

where M_i is the DCM conversion ratio of the ith LED string, which is given by the following equation.

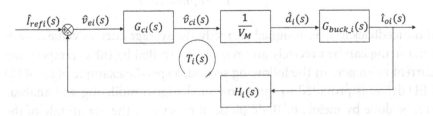

FIGURE 3.10 Small-signal block diagram of the ith string of the SIMO LED driver with closed-loop control.

$$M_i = \frac{V_{oi}}{V_{in}} = \frac{2}{1 + \sqrt{1 + \dfrac{8L_i}{mR_{LEDi}d_i^2 T_s}}} \tag{3.20}$$

A conventional PI controller is used as the compensator. The transfer function of the compensator of the ith LED string, represented by $G_{ci}(s)$, can be expressed as

$$G_{ci}(s) = \frac{sk_{p\_i} + k_{int\_i}}{s} \tag{3.21}$$

where k_{p_i} is the proportional gain, and k_{int_i} is the integral gain.

In Figure 3.10, V_M is the amplitude of the sawtooth carrier waveform. $H_i(s)$ is the sensing gain of the ith string. The output of PI compensator is fed into the PWM modulator with a gain of $1/V_M$ in order to generate the on-time duty ratio d_i. The averaged current in each LED string is determined by the corresponding current reference value $I_{refi}(s)$.

The open-loop transfer function (or loop gain) of the system, $T_i(s)$, can be written as

$$T_i(s) = G_{ci}(s) \times \frac{1}{V_M} \times G_{buck\_i}(s) \times H_i(s) \tag{3.22}$$

By substituting equations (3.19) and (3.21) into (3.22), the open-loop transfer function can be reexpressed as

$$T_i(s) = \frac{sk_{p\_i} + k_{int\_i}}{s} \times \frac{\dfrac{2V_{oi}}{d_i} \times \dfrac{1 - M_i}{2 - M_i} \times \dfrac{1}{mR_{LEDi}}}{1 + \dfrac{s}{\dfrac{2 - M_i}{(1 - M_i)mR_{LEDi}C_{oi}}}} \times \frac{1}{V_M} \times H_i(s) \tag{3.23}$$

If the feedback loop is designed correctly, the average current value in each LED string can be precisely and reliably controlled by the corresponding current reference. In the following section, a specific example of an SITO LED driver is provided to show how small-signal modeling and analysis can be done by means of Bode plots. In general, if the magnitude of the open-loop gain $|T(s)|$ is sufficiently large, the average current in each LED string should track very closely the corresponding current reference. If the *same* current reference is applied to all the LED strings, the DC current values across all strings will be the same. Hence, current balancing for multiple LED strings can be achieved. In particular, LED dimming can be enabled in any particular string by applying a smaller current reference in order to reduce the current across that string.

3.4.4 PFC

Another key design consideration of an AC-DC converter is PFC. Due to the inherent property of the time-multiplexing control scheme, only one output is physically connected to the power stage of the SITO AC-DC buck converter at any time instance. In other words, the three outputs of the SITO converter operating in DCM are fully decoupled in time. This enables us to effectively model the original SITO buck converter as three separate SISO buck converters. Let us consider the original circuit model of an ideal buck converter in Figure 3.11. $\overline{V_g}(t)$ is the average value of the pulsating DC source (i.e., the output voltage of the diode bridge).

The equivalent circuit model of the buck converter operating in DCM can be obtained by using the *average* switch network as depicted in Figure 3.12.

By operating the buck converter in DCM, the low-frequency components at the input of the switch network obeys Ohm's law [17]. Simply put, the input port of the buck converter sees an effective resistance of R_e, which is a function of the on-time duty ratio (d). The effective resistance $R_e(d)$ seen by the input port is given by

$$R_e(d) = \frac{2L}{d^2 T_s} \tag{3.24}$$

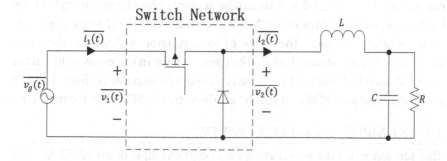

FIGURE 3.11   Circuit model of an ideal buck converter (with the switch network highlighted).

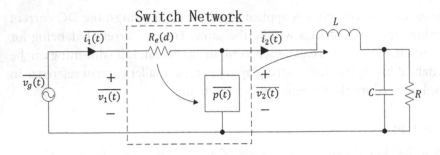

FIGURE 3.12   Equivalent circuit model of a buck converter operating in DCM.

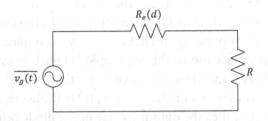

FIGURE 3.13   DC equivalent circuit model of a buck converter operating in DCM.

where d is the on-time duty ratio, L is the inductor value, and T_s is the switching period.

It is interesting to note that $R_e(d)$ is independent of the load resistance R. At low frequency, the inductor can be approximated by a short circuit. Likewise, the capacitor can be approximated by an open circuit. Hence, the buck converter can essentially be reduced to a simple voltage divider circuit shown in Figure 3.13.

The effective resistance seen by the input voltage, i.e., the sum of $R_e(d)$ and R, is purely resistive. Therefore, a high PF can be achieved even without using a PFC controller. This can be accomplished by ensuring that the buck converter operates in DCM. In the case of an SITO buck converter with unbalanced loads, the value of load resistor will *not* be the same among the three outputs. Hence, the fixed resistor in the equivalent circuit model shown in Figure 3.13 can be replaced by a variable resistor. Despite such modification, PFC can still be attained by the SITO buck converter.

3.5 EXAMPLE: A SITO LED DRIVER

This section will take you through a design example of an SITO AC-DC LED driver whose design specifications are given in Table 3.2. This LED driver is used to drive three parallel LED strings. The first, second, and

TABLE 3.2 Design Specifications

Design Parameter	Value	Design Parameter	Value
Input voltage (V_{ac})	110 V	Rated LED current (I_{LED})	350 mA
EMI filter (L_f, C_f)	1 mH, 0.1 µF	Voltage ripple factor (k)	7%
Filter capacitor (C_d)	22 µF	Sensing resistor (R_s)	1 Ω
Switching frequency (f_s)	75 kHz	Power inductor (L)	5 µH
Maximum current ripple (Δi_{L_max})	8 A	Crossover frequency (f_c)	2.5 kHz
Output capacitor (C_{o1}, C_{o2}, C_{o3})	1000 µF	Same-colored LED	String 1: 7 Blue LEDs
			String 2: 7 Blue LEDs
			String 3: 7 Blue LEDs
Rated output voltage (V_{o1}, V_{o2}, V_{o3})	String 1: 14.7 V	Distinct-colored LED	String 1: 7 Red LEDs
	String 2: 20.3 V		String 2: 7 Green LEDs
	String 3: 20.7 V		String 3: 7 Blue LEDs

third strings comprise red, green, and blue LEDs, respectively. Each string contains seven LEDs.

As a first step, the allowable range of the inductor values will be determined. By substituting the appropriate parameter values of Table 3.2 into equation (3.8), the maximum values of the inductor pertaining to the three LED strings can be obtained as $L_{1\_max} = 254$ µH, $L_{2\_max} = 336$ µH, and $L_{3\_max} = 341$ µH, respectively. From equation (3.9), the maximum theoretical value of the inductor is determined by the *smallest* of the three calculated values, i.e., $L < 254$ µH. Likewise, by substituting the relevant parameter values into equation (3.15), the minimum values of the inductor associated with the three strings can be obtained as: $L_{1\_min} = 3.52$ µH, $L_{2\_min} = 4.48$ µH, and $L_{3\_min} = 4.53$ µH, respectively. Based on equation (3.16), the minimum theoretical value of the inductor is determined to be the *largest* of the three calculated values, i.e., $L \geq 4.53$ µH. As a result, the acceptable range of the inductor value is given by 4.53 µH $\leq L \leq$ 254 µH. From a practical design perspective, it is preferable to minimize the size of the inductor to achieve a small form factor of the LED driver. Thus, L is chosen to be 5 µH since this inductance value is commercially available and is also closest to the lower bound.

Now we turn our focus to the output capacitor. By using equation (3.17), the minimum values of the output capacitor corresponding to the three output strings can be obtained as $C_{o1} \geq 902$ μF, $C_{o2} \geq 653$ μF, and $C_{o3} \geq 642$ μF. Hence, if the same capacitor value is used across the three strings, from equation (3.18), C_{o1}, C_{o2}, and C_{o3} are chosen to be 1000 μF.

For ease of analysis, the SITO buck LED driver can be functionally decomposed into three separate SISO buck drivers because of the time-multiplexed control scheme. The three output channels are virtually independent of each other. Thus, we can treat the feedback control of the original SITO driver as if there are three independent SISO drivers, and each of them has its own feedback control for regulating the corresponding output current. In other words, the controller (e.g., PI compensator) in each feedback loop can be separately tuned to regulate the current in a particular LED string. This makes it much easier to do the small-signal analysis. It is important to note that the main switch of the SITO driver operates at a switching frequency *three* times faster than that of the output switch. Hence, the frequency of the output switch (f_o) is 25 kHz.

For a PI controller, we need to derive the proper values of the proportionality constant (k_p) and integral constant (k_i) to ensure a conditionally stable closed-loop system. For ease of discussion, the first string consisting of red LEDs will be considered. Suppose an AC line voltage (V_{ac}) powering the SITO AC-DC driver has an RMS value of 110 V and a frequency of 60 Hz. The peak value of V_{ac} is $110\sqrt{2} \approx 155.56$ V. Since a 22 μF filter capacitor (C_d) is used to mitigate the double-line frequency component in the output voltage of the full-bridge rectifier, we can safely assume that the rectified voltage, which becomes the input voltage of the buck converter, resembles a pure DC signal whose value can be approximated by the peak value of V_{ac} minus the voltage drop across the two diodes of a full-bridge rectifier. From Table 3.2, the first string has an output voltage of 14.7 V and a rated current of 350 mA. To start with, we need to figure out the feedback gain $H_1(s)$. A 1-Ω resistor is used as the current sense resistor R_{s1}. (Note: To reduce the conduction loss and increase the power efficiency, a current sense resistor with a lower value, e.g., 100-mΩ or smaller, is actually preferable). The voltage across the current sense resistor will be amplified by ten times via an op amp circuit to generate a feedback voltage (V_{fb1}), which is subsequently compared with a fixed current reference I_{ref1}. Hence, the feedback gain function $H_1(s)$ can be expressed as

$$H_1(s) = \frac{V_{s1}}{i_{o1}} \times \frac{V_{fb1}}{V_{s1}} = R_{s1} \times k \qquad (3.25)$$

where i_{o1} is the current across the first string, R_{s1} is the value of the current sense resistor in the first string, and k is the gain of the op amp.

Since $R_{s1} = 1$ and $k = 10$, $H_1(s)$ is therefore equal to 10, which is frequency-independent. By substituting the relevant parameters of Table 3.2 into equation (3.22), the open-loop transfer function of the *uncompensated* system (i.e., $G_{c1}(s) = 1$) can be written as

$$T_{u1}(s) = \frac{1}{V_M} \times G_{buck\_1}(s) \times H_1(s) = \frac{56}{\frac{s}{75}+1} \qquad (3.26)$$

By setting $|T_{u1}(j\omega)| = 1$ (0 dB), the crossover frequency f_{cu1} of the *uncompensated* loop gain $T_{u1}(s)$ can be obtained as $f_{cu1} = 0.668\,\text{kHz}$. According to the design specifications in Table 3.2, the switching frequency (f_s) of the SITO converter is 75 kHz. It should be noted that the switching frequency here refers to the switching frequency of the main switch of the SITO converter. Because of the time-multiplexing control scheme, the switching frequency of the main switch (f_s) is *three* times that of the output switch. Hence, the frequency of the output switch (f_o) is 25 kHz. The desired crossover frequency of the *compensated* loop gain $T_1(s)$ is chosen to be one-tenth of the frequency of the output switch, i.e., $f_{c1} = (1/10) \times f_o = 2.5\,\text{kHz}$. At 2.5 kHz, the magnitude of the uncompensated loop gain $T'_{u1}(s)$ is given by

$$|T_{u1}(j \times 2\pi \times 2.5k)| = \left| \frac{56}{\frac{j \times 2\pi \times 2.5k}{75}+1} \right| = -11.46\,\text{dB} \qquad (3.27)$$

To obtain a unity gain at the crossover frequency of 2.5 kHz, the compensator needs to provide the uncompensated loop with a gain boost of 11.46 dB. From equation (3.21), we can write

$$|G_{c1}(j \times 2\pi \times 2.5k)| = \left| \frac{j \times 2\pi \times 2.5k \times k_{p\_1} + k_{int\_1}}{j \times 2\pi \times 2.5k} \right| = 11.46\,\text{dB} \qquad (3.28)$$

where k_{p_1} and k_{int_1} are the proportional gain and integral gain of the PI controller for the first output string.

By choosing $k_{p\_1} = 3.5$, the value of k_{int_1} can be calculated to be 20755. Thus, the transfer function of the PI controller is derived as follows.

$$G_{c1}(s) = \frac{sk_{p\_1} + k_{int\_1}}{s} = 3.5 + \frac{20755}{s} \quad (3.29)$$

Finally, by using equation (3.22), the transfer function of the open-loop gain $T_1(s)$ can be expressed in the following form.

$$T_1(s) = \left(\frac{4200}{s+75} \right)\left(\frac{3.5s + 20755}{s} \right) \quad (3.30)$$

Equation (3.30) indicates that the open loop contains a dominant pole, a low-frequency pole, and a higher-frequency zero. Figure 3.14 shows the Bode magnitude and phase plots of the three transfer functions, namely, the transfer function of the *uncompensated* loop gain, the transfer function of the *compensated* loop gain, and the transfer function of the PI controller (compensator).

Figure 3.14 clearly shows that at the unity-gain frequency, the phase margin is 70°, which means that the system is stable at the nominal operating condition. Strictly speaking, the stability analysis should also be performed at the two extreme load conditions across the entire load range, namely, the light-load condition and the heavy-load condition. As a rule of thumb, if the phase margin at either condition is at least 45°, then we can claim that the closed-loop system is conditionally stable. In addition, the Bode magnitude plot (refer to the top plot in Figure 3.14) shows that the DC gain is large. This means that the steady-state error of the LED current will be sufficiently small. Last not but least, the small-signal analysis should also be carried out for the feedback loops associated with the green and blue LED strings, respectively. We leave these as an exercise for the interested reader.

3.6 SIMULATION VERIFICATION

To verify the functionality of the SITO AC-DC buck LED driver, time-domain simulations are performed using the PSIM software. The design specifications of this SITO driver are provided in Table 3.2. The SITO driver can supply a total of three parallel LED strings. At first, the balanced load condition is investigated. The first simulation is performed based on the 220 V 50 Hz AC power supply, whereas the second one is performed using the 110 V 60 Hz AC power supply. In either case, the nominal current across all three LED strings should be the same. Figure 3.15 shows the simulated waveforms of the AC line voltage and the input current with 220 V 50 Hz AC power source. The input current is sinusoidal-like with a

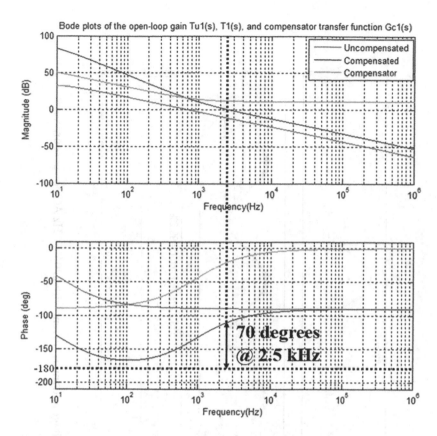

FIGURE 3.14   Bode plots of the transfer functions of the uncompensated loop gain, compensated loop gain, and the PI controller.

small distortion. Basically, the AC voltage and current are largely in phase with each other. The simulated PF is 0.992. Figure 3.16 shows the simulated waveforms of the AC line voltage and the input current with 110 V 60 Hz AC power source. Like the previous case, the AC line voltage and the input current are nearly in phase with each other. The simulated PF is 0.994. The simulation results confirm that the SITO LED driver operating in DCM yields a high PF, regardless of the supply voltage.

Figure 3.17 shows a full view of the simulated inductor current and the output branch currents of the SITO LED driver across several line cycles, assuming a 220 V 50 Hz AC source. It is interesting to note that the inductor current contains both the double-line frequency (100 Hz) component and the switching frequency (75 kHz) component. Basically, the triangular wave of the inductor current at the switching frequency is being

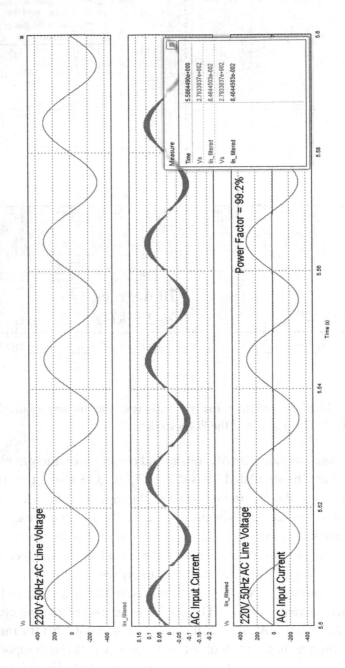

FIGURE 3.15   Simulated waveforms of the AC line voltage and input current with 220 V 50 Hz AC power source.

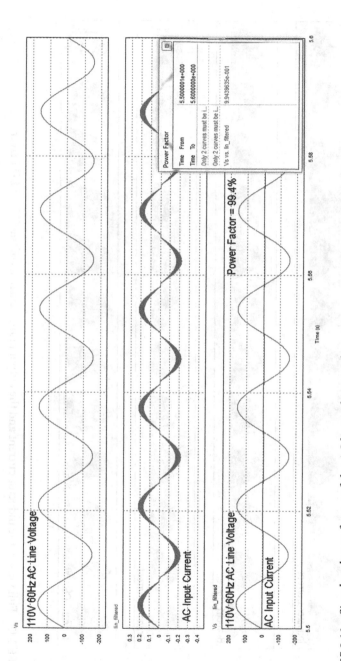

FIGURE 3.16 Simulated waveforms of the AC line voltage and input current with 110 V 60 Hz AC power source.

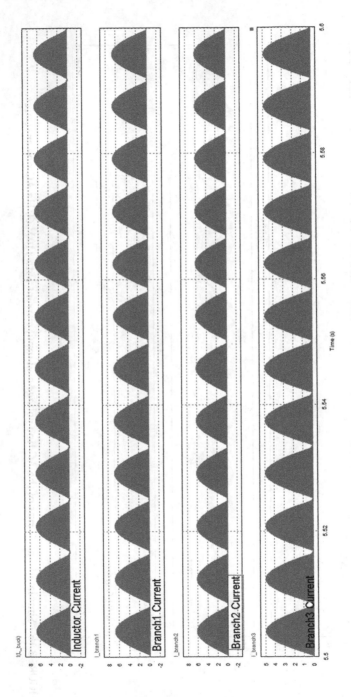

FIGURE 3.17   Full view of the inductor current and the output branch currents of the SITO LED driver.

modulated by a much lower double-line frequency. This is mainly attributed to the absence of the large DC-link capacitor in the AC-DC SITO converter, which would otherwise absorb the double-line-frequency ripple power. As a consequence, the double-line current ripple will propagate to the outputs of the SITO converter via the main inductor. The simulation results also verify the presence of the double-line frequency component at each output branch current.

Figure 3.18 shows an expanded view of the simulated waveforms of the inductor current along with the three output branch currents across a few switching cycles. It clearly shows that the SITO driver operates correctly in DCM as the inductor current returns to zero value in every switching cycle.

By setting the same current reference across the three strings, balanced load condition can be achieved. Figure 3.19 shows the simulated waveforms of the LED current across the three strings under balanced load condition. The simulation results confirm that the average current is around 350 mA in each of the three LED strings, which agrees with the target value. It should be pointed out that the scale of the vertical axis (y-axis) in Figure 3.19 is expanded. For example, in the top waveform of Figure 3.19, the resolution of the y-axis is only 5 mV. So, the AC ripple of the DC current waveform is enlarged. The percentages of the peak-to-peak value of the AC ripple over the average value of the current for the first, second, and third string are 5.43%, 6.72%, and 10.72%, respectively.

By applying distinct current references in the feedback controller, the average current values across the three LED strings will be different. This unbalanced load condition is particularly useful for color-mixing and dimming applications. Figure 3.20 shows the simulated waveforms of the three LED currents under steady-state unbalanced load condition. The average current values in the first, second, and third string are 150, 350, and 550 mA, respectively. Due to the increased resolution of the y-axis in Figure 3.20, the AC ripple of the DC current waveform is deliberately enlarged. The percentages of the peak-to-peak value of the AC ripple over the average value of the current for the first, second, and third string are 6.14%, 7.58%, and 11.28%, respectively.

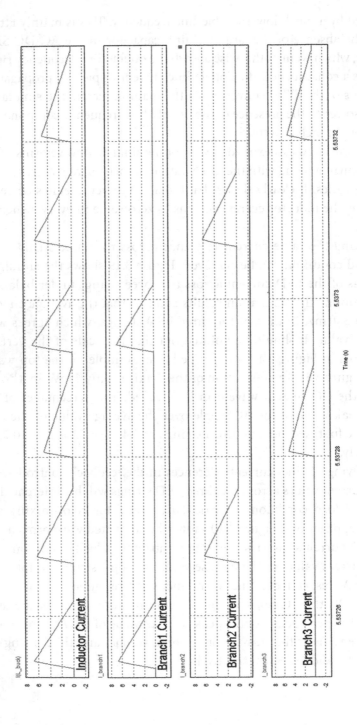

FIGURE 3.18   Expanded view of the simulated waveforms of the inductor current and the three output branch currents.

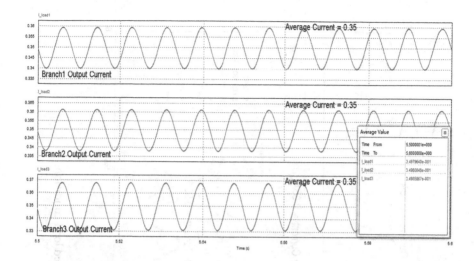

FIGURE 3.19 Simulated waveforms of LED current in the three strings in balanced load condition.

3.7 EXPERIMENTAL VERIFICATION

A hardware prototype of the single-stage AC-DC SITO LED driver is constructed for experimental verification. A photograph of this prototype is shown in Figure 3.21.

Table 3.3 contains a list of components and their respective part numbers used in the experimental prototype. The experiments require two types of LED loads. In the first experiment, same-colored LEDs are used across the three strings. Each string consists of seven blue LEDs connected in series. In the second experiment, three different-colored LEDs are used across the three strings. Specifically, the first, second, and third string consist of seven red LEDs, seven green LEDs, and seven blue LEDs, respectively. The closed-loop control enables accurate tracking of the actual LED current with the corresponding reference value. Hence, the average current value in a particular string can be independently controlled without affecting those in the other strings.

3.7.1 Basic Functionality

As a starting point, the basic functionality of the nonisolated AC-DC SITO LED driver is experimentally verified. In the first experiment, the balanced load condition is tested. Each of the three LED strings contains seven blue-colored LEDs connected in series, and the same current

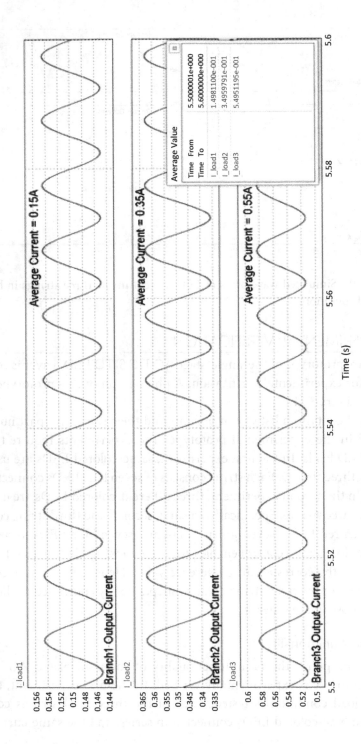

FIGURE 3.20   Simulated waveforms of LED current in the three strings in unbalanced load condition.

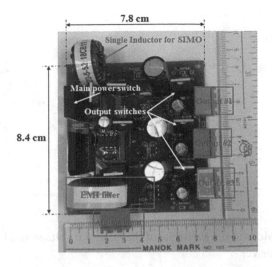

FIGURE 3.21 Photograph of the hardware prototype of the single-stage AC-DC SITO LED driver.

TABLE 3.3 List of Components Used in the Prototype

Component	Part Number	Component	Part Number
Diode bridge rectifier	GBU10G-BP	MUX	CD74HC4051E
Main switch (S_a)	IPW50R280CE	Comparator	AD8561ANZ
MOSFET gate driver	IRS2101PBF	Oscillator	LM555CN/NOPB
Freewheeling and Branch diodes	MUR1540G	Operational amplifier	OP340PA
Output switches (S_1, S_2, S_3)	IRFI4227PbF	Output capacitor (C_{o1}, C_{o2}, C_{o3})	UPX1V102MHD (long lifetime)

reference of 350 mA is applied across the three strings. Figure 3.22 shows the measured waveforms of the AC input (line) voltage and the input current of the SITO LED driver with 110 V 60 Hz AC power source and a rated output power of 30 W. It is observed that the AC line voltage and the input current are virtually in phase with each other. The measured waveform of the sinusoidal-like input current is relatively clean and smooth with some distortion at the zero crossing. The measured PF is 0.9901, which is very close to 1. Hence, the experimental results show that a high PF is achieved.

Figure 3.23 shows the full view of the measured waveforms of the inductor current (I_L) and the three output branch currents ($I_{branch1}$–$I_{branch3}$). The *absolute* peak value of the inductor current (and also the branch current)

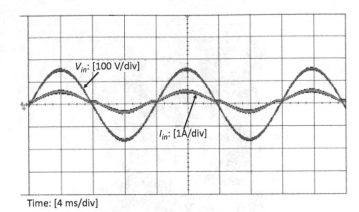

FIGURE 3.22 Measured waveforms of the AC input voltage and input current of the SITO LED driver.

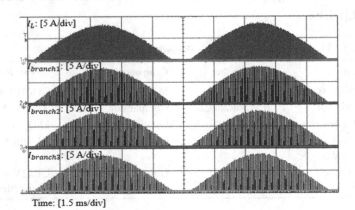

FIGURE 3.23 Measured waveforms of the inductor current and the three output branch currents in balanced load condition.

is measured to be around 7.5 A, which is within the specification limit (i.e., $\Delta i_{L\_max} = 8$ A). This helps prevent the components of the hardware proto-type from experiencing too much current stress.

Figure 3.24 show the close-up view of the gate drive signals of the main switch (V_{drive_main}) and the three output switches (V_{drive_1}, V_{drive_2}, V_{drive_3}) alongside the inductor current I_L and the three output branch currents ($I_{branch1}$, $I_{branch2}$, $I_{branch3}$) with same-colored LEDs under the balanced load condition. The experimental results show that the on-time duty ratios of the gate drive signal of the main switch (V_{drive_main}) are the same across the

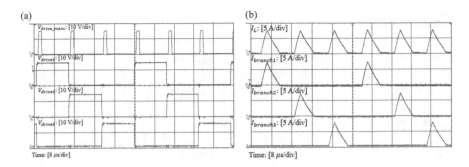

FIGURE 3.24 Close-up view of the measured waveforms of (a) the gate drive signals of the main switch (S_a) and the three output switches (S_1, S_2, S_3); (b) the inductor current (IL), and the branch currents ($I_{branch1}$, $I_{branch2}$, $I_{branch3}$) with the blue LEDs under current-balancing condition.

three LED strings. This leads to uniform peak values of I_L across switching cycles and hence, the same peak values of $I_{branch1}$, $I_{branch2}$, and $I_{branch3}$. This means that the inductor delivers *equal* current to each of the three strings. Besides, the main switch (S_a) is turned ON once every switching cycle, whereas each output switch (S_1, S_2 or S_3) is turned ON once every three switching cycles. In other words, the frequency of the main switch is *three* times that of the output switch. Consequently, the inductor current ramps up and down in every switching cycle, but each branch current only shows up once every three switching cycles. This is consistent with our expectation that the inductor current is delivered sequentially to each of the three strings in a round-robin fashion. Hence, the switching sequence of the SITO LED driver is experimentally verified. In short, the experimental results corroborate the basic functionality of the SITO topology with the time-multiplexed control scheme.

In the second experiment, the SITO LED driver is configured as an RGB LED driver whereby different-colored LEDs are used across the three LED strings. Specifically, the first, second and third LED strings are connected to red, green, and blue LEDs, respectively. To maintain current balancing, a common current reference value of 350 mA is applied across all strings. Figure 3.25 shows the close-up view of the gate drive signals of the main switch V_{drive_main} and the three output switches V_{drive_1}, V_{drive_2}, and V_{drive_3} along with I_L, $I_{branch1}$, $I_{branch2}$, and $I_{branch3}$ under the unbalanced load condition. It is observed that the on-time duty ratios of the gate drive signal of the main switch (V_{drive_main}) pertaining to the three LED channels are

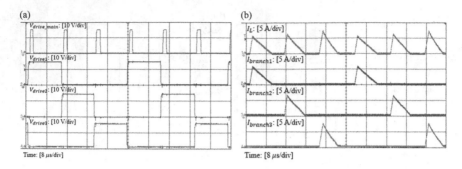

FIGURE 3.25 Close-up view of the measured waveforms of (a) the gate drive signals of the main switch (S_a) and the three output switches (S_1, S_2, S_3); (b) the inductor current (I_L) and the branch currents ($I_{branch1}$, $I_{branch2}$, $I_{branch3}$) with RGB LEDs under the unbalanced load condition.

different, which result in *distinct* peak values of I_L. Most notably, the rising and falling slopes of I_L are also different among the three channels. This can be explained by the fact that the red, green, and blue LEDs have different I-V characteristics. Even though the three different types of LEDs are biased at the same current value, the corresponding output voltages are *not* the same. Thus, the three LED channels have different power ratings. The difference in the output voltage across channels results in uneven rising or falling slopes of I_L (also, $I_{branch1}$, $I_{branch2}$, or $I_{branch3}$).

In the third experiment, the blue LEDs are used across the three output channels, but they are biased at different current values by applying distinct current references. Obviously, this will give rise to an unbalanced load condition. The first, second, and third channel are biased at 250, 350, and 450 mA, respectively. Figure 3.26 shows the close-up view of the gate drive signals of V_{drive_main}, V_{drive_1}, V_{drive_2}, and V_{drive_3} together with I_L, $I_{branch1}$, $I_{branch2}$, and $I_{branch3}$. Similar to the second experiment, the gate signal of the main switch exhibits uneven on-time duty ratios. The on-time duty ratio for the first LED channel is the smallest since it has the smallest current among the three channels. Likewise, the on-time duty ratio for the third channel is the largest since it has the largest current among the three channels. The peak values of the inductor current also show a similar trend. It has the lowest peak value for the first channel and the highest peak value for the third channel.

3.7.2 Independent Current Regulation

Based on the time-multiplexed control method using PI controllers, the average current in each of the three individual LED channels (or strings) can be independently regulated for the purpose of color-mixing and dimming.

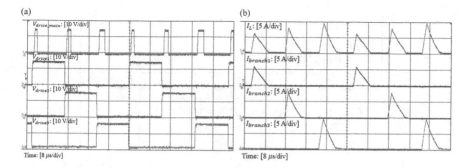

FIGURE 3.26 Close-up view of the measured waveforms of (a) the gate drive signals of the main switch (S_a) and the three output switches (S_1, S_2, S_3); (b) the inductor current (I_L) and the branch currents ($I_{branch1}$, $I_{branch2}$, $I_{branch3}$) with blue LEDs under unbalanced current condition.

To achieve uniform brightness, current balancing among the three channels is required. This can be accomplished by employing a common current reference (e.g., 350 mA) across the channels. Figure 3.27a shows the measured current waveforms of the three LED channels in steady-state condition. Clearly, it can be seen that the three individual currents lie on top of one another with the same average value of 350 mA. The peak-to-peak ripple of each LED current is within 10% of the average value of the LED. The effectiveness of the SITO LED driver in achieving current balancing among the three channels is demonstrated. On the other hand, independent current control of the output channels is also feasible by applying three distinct current references. Figure 3.27b shows the measured current waveforms of the three LED channels in steady-state condition. It shows that the average current values of the first, second, and third LED channel are 250, 350, and 450 mA, respectively. Hence, the capability of the SITO LED driver in attaining independent current regulation is demonstrated.

3.7.3 Reference Tracking

To demonstrate the effectiveness of the time-multiplexed controller, the transient response of the closed-loop system due to a step change in the current reference is experimentally verified. In this experiment, the current reference for the third LED channel is decreased almost instantaneously from 3.5 V (350 mA) to 2.5 V (250 mA). This is equivalent to a load reduction from 100% to around 70%, assuming that the rated current of the LED is 350 mA. Conversely, the current reference for the third LED channel is increased from 2.5 V (250 mA) to 3.5 V (350 mA). This corresponds to a load increase from about 70% to 100%. The current references

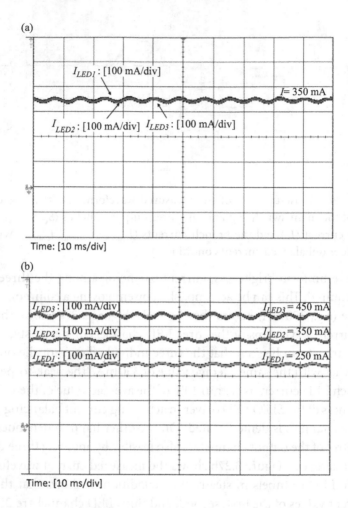

FIGURE 3.27 Measured waveforms of the three LED currents with (a) the same current reference of 350 mA and (b) distinct current references of $I_{LED1} = 250$ mA, $I_{LED2} = 350$ mA, and $I_{LED3} = 450$ mA.

of the two neighboring channels remain unchanged at 350 mA. Figure 3.28a shows the transient response of the LED currents when the closed-loop system is subject to a step-down change in the current reference of the third string. Likewise, Figure 3.28b shows the transient response of the LED currents resulting from a step-up change in the current reference of the third string.

The experimental results show that the actual value of the LED current in the third string tracks closely with its current reference. In addition, the

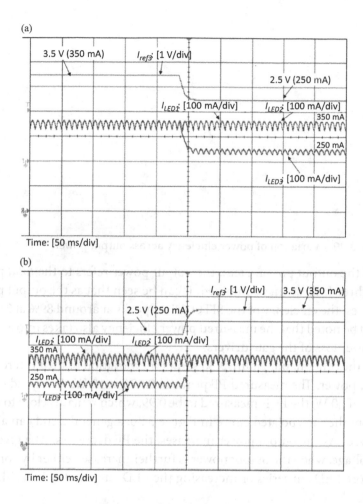

FIGURE 3.28 Transient response of the closed-loop system when it is subject to (a) a step-down change and (b) a step-up change in the current reference of the third string.

first and second LED currents remain unaffected despite a sudden change in the third LED current. No noticeable cross-regulation across the three strings is observed. The measured rise and fall times of the third LED current are approximately 25 ms.

3.7.4 Validation of Key Parameters

The key parameters of the hardware prototype of the AC-DC SITO LED driver such as power efficiency, PF, and harmonic contents of the input current are measured. Figure 3.29 shows the variation of the power efficiency

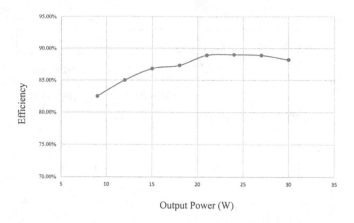

FIGURE 3.29   Variation of power efficiency across output power.

across the output power. Here, the output power refers to the *total* power of all three LED strings combined. It can be seen that as the output power increases, the efficiency of the SITO driver peaks at around 89% at 21 W. It should be noted that the measured power efficiency also takes into account the power loss of the gate drivers.

On the other hand, Figure 3.30 shows the variation of the PF across the output power. The measured PF peaks at 0.996 at 15 W. At a rated output power of 30 W, the PF is measured to be 0.99, which remains close to unity.

From the interpolated curve in Figure 3.30, a general trend can also be observed. As the output power increases, the PF decreases. At a given AC line voltage, when the output power is further increased either by connecting more LEDs in series or increasing the LED current, the PF is likely to

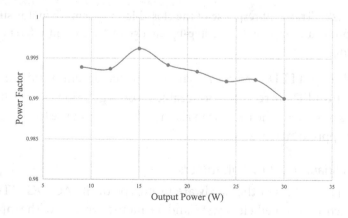

FIGURE 3.30   Variation of PF across output power.

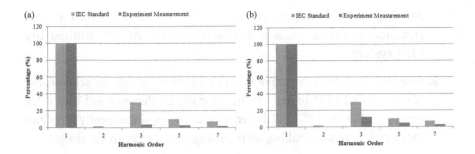

FIGURE 3.31 Comparison of the measured harmonic contents of the input current (blue color) against the corresponding harmonic current limit specified in the IEC61000-3-2 standard (orange color) at (a) 30 W rated output power and (b) 3 W output power.

drop below 0.99 due to increased distortion in the AC line current at the zero-crossing point (formally known as crossover distortion). This non-conducting interval at which the input current is zero actually depends on the output (DC) voltage of the SITO driver. Simply put, the higher the output voltage, the longer this interval will become. Therefore, when the output power is increased further, the PF is expected to decrease quite rapidly. In practice, the AC-DC SITO LED driver is designed in such a way that the PF at the rated output power is no less than 0.99.

The SITO LED driver belongs to Class C equipment under the IEC 61000-3-2 standard [15]. The harmonic contents of the input current are measured and compared against the corresponding harmonic current limit specified in the IEC61000-3-2 standard. Figure 3.31a shows the measured harmonic currents against the harmonic current limits at a rated output power of 30 W. The same measurement is repeated at a lower output power of 3 W, which is 10% of the rated output power. Figure 3.31b shows the measured harmonic currents against the harmonic current limits at 3 W output power. The experimental results clearly show that the first few odd harmonics of the measured input current are lower than the respective harmonic current limit as set forth in the IEC61000-3-2 standard.

3.8 SUMMARY OF KEY POINTS

1. An offline single-stage SIMO AC-DC LED driver employs only a single inductor in the power stage to drive multiple independent LED strings. The elimination of the intermediate high-voltage E-cap at the DC link, which is typically required in a conventional

AC-DC power supply, increases the operating lifetime and enhances the robustness and reliability of the AC-DC multistring LED system.

2. Due to the inherent nature of the time-multiplexed control scheme, the SITO AC-DC LED driver can functionally be decomposed into three separate SISO AC-DC LED driver. Such decomposition greatly simplifies the circuit analysis, small-signal modeling, and the controller design.

3. Both simulation and experimental results demonstrate the effectiveness of the AC-DC SITO LED driver in achieving fully independent current control of each individual LED string with no noticeable cross-regulation. The average current of each string can be precisely controlled by setting an appropriate current reference. No additional current-balancing circuitry is required.

4. By operating the SIMO AC-DC LED driver in DCM, high PF can be achieved without using a PFC controller. The experimental results show that the measured PF is 0.99 at the rated output power of 30 W.

5. The novel SIMO AC-DC topology carries the key advantages of a smaller component count, lower BOM cost, simplified control scheme, ease of implementation, better scalability to multiple outputs, and high efficiency.

REFERENCES

1. Y. Guo, S. Li, A. T. L. Lee, S.-C. Tan, C. K. Lee, and S. Y. R. Hui, "Single-stage AC/DC single-inductor multiple-output LED drivers", *IEEE Trans. Power Electron.*, vol. 31, no. 8, pp. 5837–5850, Aug. 2016.
2. A. T. L. Lee, W. Jin, S. Li, S. C. Tan, and S. Y. R. Hui, "Single-stage single-inductor multiple-output (SIMO) inverter topology with precise and independent amplitude control for each AC output", U.S. Patent 10903756 B2, Jan. 26, 2021.
3. D. Ma, W. H. Ki, P. K. T. Mok, and C. Y. Tsui, "Single-inductor multiple-output switching converters with bipolar outputs", *Proceedings - IEEE International Symposium on Circuits and Systems, 2001*, pp. 301–304, May 2001.
4. D. Ma, W.-H. Ki, C.-Y. Tsui, and P. K. T. Mok, "Single-inductor multiple-output switching converters with time- multiplexing control in discontinuous conduction mode," *IEEE J. Solid-State Circuits*, vol. 38, no. 1, pp. 89–100, Jan. 2003.

5. D. Ma, W.-H. Ki, and C.-Y. Tsui, "A pseudo-CCM/DCM SIMO switching converter with freewheel switching", *IEEE J. Solid-State Circuits*, vol. 38, no. 6, pp. 1007–1014, Jun. 2003.
6. D. Kwon and G. A. Rincon-Mora, "Single-inductor-multiple-output switching DC-DC converters," *IEEE Trans. Circuits Syst., II: Exp. Briefs*, vol. 56, no. 8, pp. 614–618, Aug. 2009.
7. X. Jing, P. Mok, and M. Lee, "A wide-load-range constant-charge-auto-hopping control single-inductor- dual-output boost regulator with minimized cross-regulation," *IEEE J. Solid-State Circuits*, vol. 46, no. 10, pp. 2350–2362, Oct. 2011.
8. H. Chen, Y. Zhang, and D. Ma, "A SIMO parallel-string driver IC for dimmable LED backlighting with local bus voltage optimization and single time-shared regulation loop," *IEEE Trans. Power Electron.*, vol. 27, no. 1, pp. 452–462, Jan. 2012.
9. A. Lee, J. Sin, and P. Chan, "Scalability of quasi-hysteretic FSM-based digitally- controlled single-inductor dual-string buck LED driver to multiple string," *IEEE Trans. Power Electron.*, vol. 29, no. 1, pp. 501–513, Jan. 2014.
10. Y. Zhang and D. Ma, "A fast-response hybrid SIMO power converter with adaptive current compensation and minimized cross-regulation," *IEEE J. Solid-State Circuits*, vol. 49, no. 5, pp. 1242–1255, May 2014.
11. C.-W. Chen and A. Fayed, "A low-power dual-frequency SIMO buck converter topology with fully-integrated outputs and fast dynamic operation in 45 nm CMOS," *IEEE J. Solid- State Circuits*, vol. 50, no. 9, pp. 2161–2173, Sep. 2015.
12. K. Modepalli and L. Parsa, "A scalable N-color LED driver using single inductor multiple current output topology," *IEEE Trans. Power Electron.*, vol. 31, no. 5, pp. 3773–3783, May 2016.
13. S. Huynh and C. V. Pham, "Single inductor multiple LED string driver," U.S. Patent 20120043912 A1, Feb. 23, 2012.
14. H. Kim, C. Yoon, H. Ju, D. Jeong, and J. Kim, "An AC-powered, flicker-free, multi-channel LED driver with current-balancing SIMO buck topology for large area lighting applications," Proceedings IEEE Applied Power Electronics Conference and Exposition, pp. 3337–3341, 2014.
15. Electromagnetic compatibility (EMC)—Part 3: Limits-Section 2: Limits for harmonic current emissions (equipment input current ≤ 16 A per phase), IEC Standard IEC61000-3-2, 2018.
16. S. Li, S. C. Tan, C. K. Lee, E. Waffenschmidt, S. Y. R. Hui, and C. K. Tse, "A survey, classification and critical review of light-emitting-diode drivers," *IEEE Trans. Power Electron.*, vol. 31, no. 2, pp. 1503–1516, Feb. 2016.
17. R. W. Erickson and D. Maksimovic, *Fundamentals of Power Electronics*, 2nd ed. New York, NY, USA: Springer, 2001.
18. R.-L. Yin and Y.-F. Chen, "Equivalent circuit model of light-emitting-diode for system analysis of lighting drivers", *2009 IEEE Industry Applications Society Annual Meeting*, Oct. 2009.
19. LUXEON Rebel Color Line, Lumileds Holding (2020). [Online] Available: http://www.lumileds.com/products/color-leds/luxeon-rebel-color/.

PROBLEMS

3.1 Discuss the advantages of having a single-stage SIMO AC-DC LED driver when compared to its two-stage SIMO counterpart.

3.2 Under typical operating condition, a red LED chip has a threshold voltage of 0.65 V and a forward voltage of 2.05 V with a rated current of 350 mA. The green LED chip has a threshold voltage of 0.7 V and a forward voltage of 2.8 V with a rated current of 350 mA. The blue LED chip has a threshold voltage of 0.9 V and a forward voltage of 3 V with a rated current of 350 mA. The electrical properties of an LED device can generally be represented by the equivalent circuit model as illustrated in Figure 3.8.

a. Find the equivalent resistances of the red, green, and blue LED chip, respectively.

b. Calculate the rated power values of the red, green, and blue LED chip, respectively.

c. An RGB LED light strip is made up of three LED strings connected in parallel. In the first LED string, there are nine red LED chips connected in series. The second LED string comprises seven series-connected green LED chips while the third LED string comprises six series-connected blue LED chips. Determine the total output power of the three LED strings when the current in each string is 350 mA.

3.3 A conventional PI controller is used to independently regulate the current in each of the three LED strings of a SITO AC-DC LED driver. The transfer function of the PI controller is given by equation (3.21). The PI controller can be realized by a simple analog circuit consisting of an operational amplifier and resistor-capacitor network shown in Figure 3.32.

a. Derive the transfer function of the op amp circuit G(s), where $G(s) = V_o/Vi$.

b. Express k_p and k_i in terms of the resistors and capacitor of the op amp circuit shown in Figure 3.32, where k_p is the proportional gain, and k_i is the integral gain of the PI controller.

c. Determine the locations of the compensated pole and zero introduced by the PI controller.

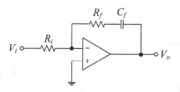

FIGURE 3.32   Realization of a PI controller using an inverting op amp circuit.

d. Design the op amp circuit for the PI controller with $k_p = 0.01$ and an integral break frequency of 100 rad/sec. Assume $R_i = 100$ kΩ.

In Problems 3.4–3.7, consider an ideal single-inductor three-string AC-DC LED driver with the following design specifications:

RMS value of the AC input voltage	$V_{ac,\,RMS} = 220$ V
Line frequency	$f_{line} = 50$ Hz
EMI filter (L_f, C_f)	$L_f = 1$ mH, $C_f = 0.1$ μF
Filter capacitor (after the full-bridge rectifier)	$C_d = 22$ μF
Switching frequency of the main switch	$f_s = 75$ kHz
Current sensing resistor per LED string	$R_{sense} = 100$ mΩ
Maximum allowable inductor current ripple	$\Delta I_{L,\,max} = 8$ A
Maximum output voltage ripple (%)	$\Delta V_o = 8\%$
Feedback (sensing) gain per LED string	$H(s) = 10$
Peak amplitude of the sawtooth carrier	$V_M = 5$

For red LED:

Rated current	$I_R = 350$ mA
Forward voltage per LED	$V_{F,R} = 2.05$ V (at 350 mA)
Threshold voltage per LED	$V_{th,\,R} = 0.65$ V
Total number of red LEDs in the first string	$N_R = 7$ (connected in series)

For green LED:

Rated current	$I_G = 350$ mA
Forward voltage per LED	$V_{F,G} = 2.8$ V (at 350 mA)
Threshold voltage of per LED	$V_{th,\,G} = 0.7$ V
Total number of red LEDs in the second string	$N_G = 7$ (connected in series)

For blue LED:

Rated current	$I_B = 350\,\text{mA}$
Forward voltage per LED	$V_{F,B} = 3.0\,\text{V}$ (at 350 mA)
Threshold voltage per LED	$V_{th,B} = 0.9\,\text{V}$
Total number of red LEDs in the second string	$N_B = 7$ (connected in series)

All losses can be ignored.

3.4

a. Determine the allowable range of the inductor value.

b. Based upon the result in part (a), choose a practical value of the inductor. Explain your choice.

c. Find the minimum value of the output capacitor.

3.5 A time-multiplexed control scheme using PI compensators as shown in Figure 3.4 is used to achieve independent current regulation of each LED string of the single-inductor three-string AC-DC LED driver. Consider the feedback loop for the red LED string in this problem.

a. Derive the open-loop transfer function of the uncompensated system (i.e., the system *without* PI compensation).

b. Suppose the desired crossover frequency of the compensated loop gain (the open-loop gain *with* compensation) is chosen to be around one-tenth of the output switching frequency. In addition, the proportional gain of the PI compensator (k_p) is chosen to be 3.5. Obtain the transfer function of the PI compensator.

c. Determine the loop gain function (i.e., the open-loop transfer function of the compensated system).

d. Generate the Bode plots of the loop gain from part (c). Obtain the phase margin.

3.6

   a. Design the PI compensators for the feedback loops pertaining to the green LED string and the blue LED string, respectively. Show that the two closed-loop systems are stable.

   b. Construct a simulation model for the single-inductor three-string AC-DC LED driver with closed-loop control based on the aforementioned specifications. Use a simulation software (e.g., Pspice, PSIM, Simulink, etc.) to plot the steady-state waveforms of the gate-drive signals of the main switch and the three output switches, the inductor current, and the three LED currents.

   c. Based on the simulation results in part (b), obtain the average values of the three LED currents. Show how effective your closed-loop system is in attaining precise and independent current regulation across the three LED strings. Also, verify that the requirement of the maximum inductor current ripple of 8 A is satisfied.

3.7 To achieve smaller size and lower cost, it is desirable to replace the 22 μF film capacitor (C_d) at the output of the full-bridge rectifier by a 0.1 μF ceramic capacitor with the same maximum voltage rating.

   a. Explain why the closed-loop system is no longer stable when the value of C_d is reduced to 0.1 μF, assuming the other parameter values remain unchanged.

   b. Redesign the PI compensators to stabilize the system. Derive the transfer functions of the new PI compensators for the red, green, and blue LED strings, respectively.

   c. Based on your method in part (b), update your simulation model from Problem 3.6 part (b) accordingly. Use simulation to verify the stability of the closed-loop system.

CHAPTER 4

Single-Inductor Multiple-Output DC-AC Boost Inverters

4.1 INTRODUCTION

An inverter is an electronic device or circuitry that receives power from a direct current (DC) power source, such as a battery or solar panel, and converts it to an alternating current (AC) supply for feeding inductive loads. Opposite to that of a rectifier, the main function of an inverter is to perform DC to AC power conversion. In general, there are three major types of inverters, namely, pure sine wave (also known as true sine wave), modified sine wave, and square wave. Owing to the fact that the pure sine wave inverters can generate clean sinusoidal AC voltages with low harmonic distortion, they are more suitable for powering household appliances and various electronic devices for achieving higher efficiency, increased robustness, and reliability.

The single-inductor multiple-output (SIMO) DC-DC and AC-DC topologies have been elucidated in previous chapters. It is therefore natural to extend the concept of SIMO to DC-AC power conversion. In this chapter, a new nonisolated DC-AC (inverter) topology known as the SIMO DC-AC boost inverter is presented [1]. In simple terms, this inverter transforms a single DC input supply into multiple independent sinusoidal-like AC

DOI: 10.1201/9781003239833-4

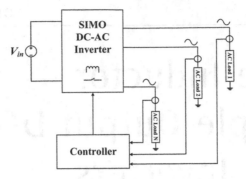

FIGURE 4.1   Overview of the SIMO DC-AC boost inverter with feedback control.

outputs by employing only a single inductor in the power stage. The root-mean-square (RMS) value of the AC output voltage is higher than that of the DC input voltage due to the inherent boost (step-up) operation. As will be explained shortly, this SIMO-based inverter is capable of producing sinusoidal-like output voltages with low distortion. Hence, it can be treated as a pure sine wave inverter. Figure 4.1 gives an overview of the SIMO DC-AC boost inverter whose AC outputs are regulated by a single feedback controller.

The SIMO-based inverters can be used in various practical applications requiring multiple independently controlled AC power sources. For instance, they can deliver AC power sequentially to multiple transmitting coils in a multi-coil wireless power transfer (MC-WPT) system in a round-robin time-multiplexed manner. A conventional multi-coil wireless power transmitter typically employs separate power inverters, each of which drives a transmitting (primary) coil that is closely coupled with the corresponding receiving (secondary) coil [2–6]. Perhaps, one of the most common transmitter topologies for a simple two-coil WPT system is a full-bridge resonant inverter, which consists of a full-bridge inverter connected to a series LC resonant tank [6]. First, the full-bridge inverter generates a square waveform that contains a lot of harmonics. Subsequently, the LC resonant circuit acts as an output filter by suppressing the high-order harmonics in order to produce a smooth sinusoidal-like AC output voltage with low distortion. Figure 4.2 shows the ideal circuit topology of the conventional MC-WPT system using multiple full-bridge inverters. Even though this particular topology looks modular, its major drawback is that the number of full-bridge inverters is directly proportional to the number of transmitting coils, which leads to a larger form factor, lower

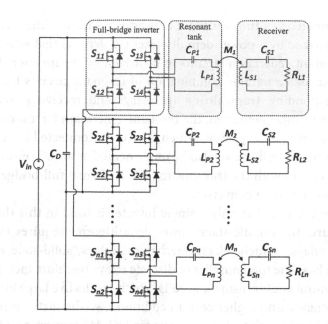

FIGURE 4.2 Simplified circuit topology of a conventional MC-WPT system using multiple full-bridge inverters.

efficiency, and higher cost, especially when a larger number of transmitting coils are required.

Alternatively, a three-stage DC-AC power conversion architecture for driving multiple transmitting coils has been reported in [6–11], as shown in Figure 4.3. The first stage is a DC-AC converter (inverter), which can

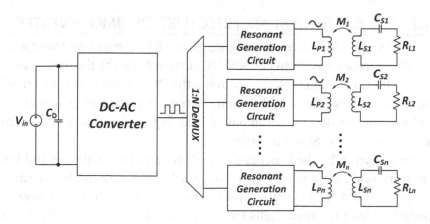

FIGURE 4.3 System architecture of a three-stage DC-AC power conversion for MC-WPT.

be implemented as a full-bridge (or half-bridge) inverter. The second stage is characterized by a power demultiplexer (or demux) that selects one or more resonant generation circuits to be driven by the inverter. The third stage comprises a parallel combination of resonant circuits loaded with the corresponding transmitting (primary) and receiving (secondary) coils. In real implementation, the resonant generation circuit can consist of just one resonant capacitor, which is directly connected in series with the transmitting coil. It can also be implemented as an LC resonant tank, which combines with the transmitting coil to form a full-bridge (or half-bridge) LLC resonant converter.

Despite the fact that only a single inverter is used in this three-stage architecture, the middle-stage power demultiplexer requires the use of multiple relays, namely, electromechanical relays, solid-state relays, or FET switches. The total number of discrete relays therefore increases with a greater number of transmitting coils, which results in a large form factor, lower efficiency, and higher cost. In addition, two distinct controllers are employed in this architecture, i.e., one for DC-AC converter and another for the power demultiplexer. A larger number of transmitting coils further complicates the controller design. The two controllers are also required to be well synchronized with each other for proper power flow management. In view of such issues, a single-stage SIMO inverter is introduced which is capable of achieving a compact, efficient, scalable, and cost-effective MC-WPT system. The general system architecture of the SIMO inverter will be introduced in the following section.

4.2 GENERAL SYSTEM ARCHITECTURE OF SIMO INVERTER

To achieve DC-AC power conversion, the SIMO inverter first transforms the DC voltage source into periodic DC current pulses by the main inductor. Each LC resonant tank then converts the DC current pulse into an AC power source oscillating at the resonant frequency. Each of these individual AC power sources drive an output load. Figure 4.4 depicts the system architecture of the SIMO inverter.

Fundamentally, the functions of a SIMO DC-DC converter and the resonant tank, acting as a DC-AC stage, are effectively combined into a single stage in a SIMO inverter. The key components of the SIMO inverter include at least the main inductor, power switches (e.g., MOSFETs), and a parallel combination of LC resonant tanks. The main inductor serves as an ideal current source that delivers the storage energy sequentially

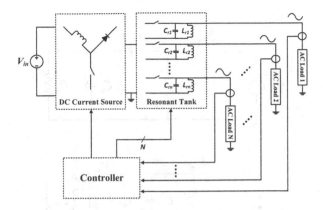

FIGURE 4.4   System architecture of the SIMO inverter.

to each of the AC outputs in a round-robin time-multiplexed manner. Due to the finite storage energy of the main inductor, there exists a theoretical maximum achievable output power for each AC output, which will be analytically derived later. Ideally, each individual resonant tank performs AC excitation by converting the periodic current pulses from the main inductor into a pure sinusoidal output voltage. The controller is primarily responsible for generating the proper switching sequence of all switches. It also regulates the peak (or RMS) value of the sinusoidal output voltage by determining the on-time duty ratio of the main switch corresponding to each output. In this architecture, only a single controller is needed to simultaneously regulate all the AC output voltages. Hence, the control scheme is simple and scalable, even with an increasing number of AC loads. Compared with the existing multiple-output inverters, the SIMO inverter requires a fewer number of power switches, gate drivers, and passive components, which makes it highly compact, efficient, scalable, and cost-effective.

4.3 CIRCUIT TOPOLOGY

In principle, the SIMO inverter topology is formulated by integrating the functions of the conventional DC-DC switching converter such as a boost converter [see Figure 4.5a] and a parallel network of LC resonant tanks [see Figure 4.5b] into a single stage. Figure 4.5c shows the resulting circuit topology of the SIMO boost inverter. The independent current source (I_{DC}) for supplying a parallel network of LC tanks can readily be replaced by the main inductor L in the boost converter as it acts as a current source

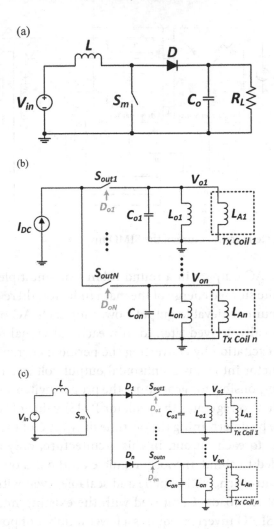

FIGURE 4.5 (a) Conventional DC-DC boost converter; (b) a parallel network of LC resonant tanks; (c) circuit topology of the SIMO boost inverter.

for delivering the storage energy successively to each of the individual AC outputs. Since the boost converter and the resonant tanks share a common ground, they can be combined easily and naturally to form a single-stage multiple-output inverter. Hence, a SIMO boost inverter with a total number of n sinusoidal-like AC outputs can be obtained. Notice that only a single inductor L is required in the power stage to drive any number of AC outputs.

Without loss of generality, a single-inductor three-output (SITO) boost inverter is considered. Figure 4.6 shows the ideal circuit diagram of the SITO boost inverter operating in discontinuous conduction mode (DCM). For ease of discussion, all circuit components of the SITO inverter are assumed to be ideal with zero losses.

Figure 4.6 shows that the SITO boost inverter transforms a single DC input voltage (V_{in}) into three independent AC output voltages (V_{o1}, V_{o2}, V_{o3}). It employs four power switches, namely, the main switch (S_{main}) and three output switches (S_{out1}, S_{out2}, S_{out3}). The power switch can be implemented using a silicon MOSFET. In Chapter 6, we will consider an alternative implementation of the boost inverter whose power switch is implemented using gallium nitride (GaN) transistor. The output branch current flowing across the three output switches is denoted by I_{o1}, I_{o2}, and I_{o3}, respectively. The symbol L represents the main inductor, and I_L represents the main inductor current. Each output of the SITO inverter is characterized by the corresponding LC resonant tank, i.e., (L_{o1}, C_{o1}), (L_{o2}, C_{o2}), or (L_{o3}, C_{o3}), which forms an integral part of the power stage, together with the corresponding inductive load, namely, L_{T1}, L_{T2} or L_{T3}. For practical applications of wireless

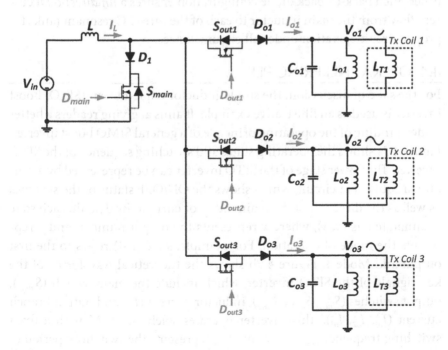

FIGURE 4.6   Ideal circuit diagram of the SITO boost inverter.

power transfer, the inductive load serves as the transmitting (primary) coil. It is also worth mentioning that the circuit of the SITO inverter contains a total of four blocking diodes, namely, D_1, D_{o1}, D_{o2}, and D_{o3}. For example, D_1 is used to prevent an unintended current flow from the common ground to any one AC output (whose instantaneous output voltage is negative) via the body diode of S_{main}. This could potentially occur in the discharging phase (or the second subinterval) of DCM when S_{main} is turned OFF and any one of the output switches is turned ON. Likewise, the blocking diodes in the output branches (D_{o1}, D_{o2}, D_{o3}) are used to prevent an unwanted reverse current flow from the AC output to the DC input via the corresponding body diodes of the output switches (D_{body_o1}, D_{body_o2}, D_{body_o3}). This could happen during the time interval when the instantaneous value of the AC output voltage becomes larger than the input DC voltage. As shown in Figure 4.6, the internal body diode of the MOSFET and the external blocking diode constitute a pair of back-to-back diode structure. For instance, the internal body diode of S_{out1} and the blocking diode D_{o1} are placed in series but pointing in opposite directions. Likewise, the internal body diode of S_{main} and the blocking diode D_1 are connected in series but pointing toward each other. Such back-to-back diode configuration ensures a *unidirectional* current flow from the main inductor to each of the three LC resonant tanks for proper circuit operation under all circumstances.

4.4 OPERATING PRINCIPLE

For the sake of discussion, the single-inductor triple-output (SITO) boost inverter is used as an illustrative example. It aims at giving readers a better understanding of the operating principle of a general SIMO boost inverter. Figure 4.7a shows the operating states and switching sequence of the SITO inverter. The power stage of the SITO inverter can be represented by a simplified equivalent circuit, which shows the ON/OFF status of the switches as well as the direction of the main inductor current. By default, each state is annotated as (x, y), where x represents the output number, and y represents the mode of operation. For example, state (1–1) refers to the first output and Mode 1. Figure 4.7b shows the theoretical waveforms of the key signals of the SITO inverter, which include the main switch (S_{main}), output switches (S_{out1}, S_{out2}, S_{out3}), inductor current (I_L), and output branch current (I_{o1}, I_{o2}, I_{o3}). This inverter operates solely in DCM with a fixed switching frequency f_{sw}. The symbol T_s represents the switching period of the inverter, where $T_s = \dfrac{1}{f_{sw}}$. The on-time duty ratios of S_{main} associated

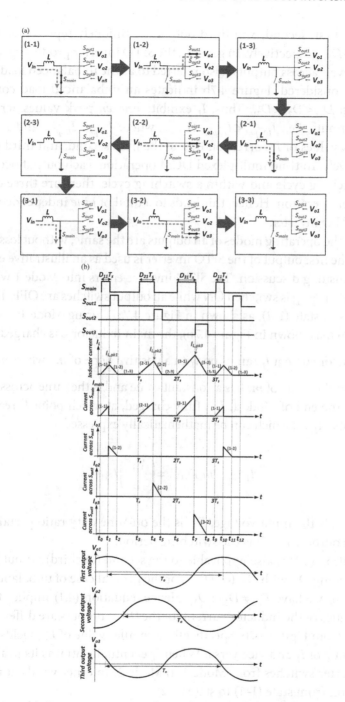

FIGURE 4.7   (a) Operating states and switching sequence of the SITO inverter; (b) theoretical waveforms of the main switch, output switches, main inductor current, and the current flowing across the main switch and the output switches.

with the first, second, and third output are uniquely represented by D_{11}, D_{21}, and D_{31}, respectively. In general, the SITO inverter producing *unequal* power levels across outputs (also referred to as the unbalanced load condition) is considered. Figure 4.7b indicates an unbalanced load condition in which $D_{11} < D_{21} < D_{31}$. Thus, I_L exhibits *uneven* peak values across the three outputs, i.e., $I_{L,\,pk1} < I_{L,\,pk2} < I_{L,\,pk3}$, where $I_{L,\,pk1}$, $I_{L,\,pk2}$, and $I_{L,\,pk3}$ are the peak values of I_L corresponding to the first, second, and third output, respectively. In time-multiplexed DCM operation, each output occupies a full switching cycle and within a switching cycle, there are three distinct modes of operation. Hence, this leads to a total of *nine* independent states of the SITO inverter.

Since the operating modes of all outputs are the same, without loss of generality, the first output of the SITO inverter is used as an illustrative example for the ensuing discussion. The SITO inverter enters into Mode 1 when the main switch S_{main} is switched ON while all output switches are OFF. This corresponds to state (1–1), as shown in Figure 4.7a. During Mode 1 (e.g., from time t_0 to t_1 as shown in Figure 4.7b), the main inductor L is charged up and the inductor current I_L ramps up with a positive slope of m_1, where $m_1 = \dfrac{V_{in}}{L}$. Note that the value of m_1 has a constant value and is the same across all outputs. At the end of Mode 1, L is fully charged, at which point I_L reaches its peak value, $I_{L,\,pk1}$, which can be mathematically expressed as

$$I_{L,pk1} = m_1 D_{11} T_s = \left(\frac{V_{in}}{L}\right) D_{11} T_s \qquad (4.1)$$

where V_{in} is the input voltage, D_{11} is the on-time duty ratio pertaining to the first output, and T_s is the switching period.

Equation (4.1) is also applicable to the second (or third) output by simply replacing D_{11} with D_{21} (or D_{31}). For the general case of unbalanced load condition, we have $D_{11} \neq D_{21} \neq D_{31}$. Hence, equation (4.1) implies that the peak values of the inductor current for the three outputs are different, as is evident from Figure 4.7b. Specifically, a smaller value of D_{11} yields a lower peak value of I_L and vice versa. When I_L eventually attains its peak value, the inverter switches from Mode 1 to Mode 2. In other words, it makes a transition from state (1–1) to state (1–2).

In Mode 2, S_{main} is switched OFF, and the first output switch S_{out1} is switched ON, while the other two output switches (S_{out2}, S_{out3}) are OFF.

This corresponds to state (1–2), as depicted in Figure 4.7a. In this mode (e.g., from time t_1 to t_2 as shown in Figure 4.7b), the energy stored in the main inductor L is released to the first AC output of the inverter. Because of the discharging of L, the inductor current I_L decreases with a negative slope of $m_2(t)$, where $m_2(t) = \dfrac{[V_{in} - V_{o1}(t)]}{L}$, and $V_{o1}(t)$ is the sinusoidal voltage of the first output. Since this inverter performs boost operation, the RMS value of $V_{o1}(t)$ must be greater than V_{in} in Mode 2, i.e., $V_{in} < V_{o1}(t)$. Hence, the value of $m_2(t)$ is always negative, which implies a falling slope of I_L. Strictly speaking, due to the sinusoidal nature of $V_{o1}(t)$, $m_2(t)$ is a nonlinear function that resembles a convex (downward) slope where its first derivative is increasing. This is in contrast with the constant (rising) slope of m_1 in Mode 1. In general, the time duration of Mode 2 is relatively short, compared to a full resonant cycle (T_o). Therefore, the variations in $V_{o1}(t)$ are reasonably small such that it can be approximated by its average value during Mode 2. As a first-order approximation, the downward slope of I_L can be largely represented by a straight line with a constant slope of m_2 [see Figure 4.7b], which is given by

$$m_2(t) = \frac{V_{in} - \overline{V_{o1,m2}}}{L} \tag{4.2}$$

where $\overline{V_{o1,m2}}$ is the average value of the first output voltage in Mode 2.

The output switch S_{o1} remains closed until I_L returns to zero value. At the end of Mode 2, L is fully discharged, and S_{o1} is switched OFF at the zero-crossing of I_L. The SITO inverter then transitions from Mode 2 to Mode 3 or, more precisely, from state (1–2) to (1–3).

In Mode 3, all switches are in their open (OFF) position, as can be seen in state (1–3) in Figure 4.7a. I_L remains at *zero* value in this mode, which is also referred to as the idle phase in DCM. Even though no useful energy transfer is performed, the presence of this idle phase is vital to prevent unwanted cross-regulation across the three outputs of the SITO inverter, which will be explained later. The end of state (1–3) marks the completion of the switching cycle of the first output and the beginning of the switching cycle of the second output.

Subsequently, the above mode transitions are also applicable to the second output with the exception that in state (2–2), S_{o2} is switched ON, while S_{o1} and S_{o3} are OFF. It should be noted that only one output switch can be turned ON per switching cycle. State (2–1) and (2–3) are virtually identical to state (1–1) and (1–3), respectively. The switching cycle of the third output follows immediately after that of the second output. Likewise, the same mode transitions are applied to the third output, except that in state (3–2), S_{o3} is switched ON, while S_{o1} and S_{o2} are OFF. At the end of state (3–3), the SITO inverter completes one full resonant cycle T_o, where $T_o = 3T_s$. Afterwards, the inverter makes a transition back to state (1–1), and the whole switching process is repeated. In essence, the energy stored in the main inductor is distributed across each individual AC output in a round-robin time-interleaving manner. The amount of energy being delivered to each AC output from the shared inductor can be independently adjusted by controlling the on-time duty ratio of the main switch S_{main} associated with that output.

The switching frequency f_{sw} of the SITO boost inverter is *three* times the resonant frequency of the LC resonant tank f_o, which can be mathematically expressed as

$$f_{sw} = 3f_o = \frac{3}{2\pi\sqrt{L_o C_o}} \tag{4.3}$$

where L_o is the resonant inductor, and C_o is the resonant capacitor.

It is assumed that the three resonant tanks at the outputs of the SITO inverter have the same values of L_o (or C_o), which implies that the same resonant frequency is produced across all outputs. For practical WPT, the resonant frequency needs to fall within the range of frequencies specified in the WPT standard. For instance, according to the Qi standard, the operating frequency ranges between 110 and 205 kHz for the low-power Qi chargers (up to 5 W), and 80 and 300 kHz for medium-power Qi charges (up to 120 W). At a given resonant frequency, the appropriate values of L_o and C_o in the resonant tank can therefore be determined.

4.5 THEORETICAL DERIVATIONS OF SINUSOIDAL OSCILLATION

One of the salient characteristics of the SITO boost inverter is the sinusoidal oscillation at the individual outputs. In this section, a complete mathematical proof is provided in order to show that the SIMO inverter is capable of producing sinusoidal-like oscillation. But first, we would like to elucidate how the SITO inverter can be decomposed into three independent single-inductor single-output (SISO) inverters. This provides the theoretical underpinning for simplifying the analysis of SITO inverter by treating it as three SISO inverters.

Because of DCM operation, the DC input voltage (V_{in}) is converted into periodic current pulses by the main inductor L, of which the stored energy is transferred successively to each of the three outputs in a time-multiplexed manner. Figure 4.8 shows the ideal waveforms of the output enable signals, the gate-drive signals of the switches, the main inductor current, and three hypothetical inductor currents corresponding to the first, second, and third output. Since the inductor current (I_L) always starts from zero value at the beginning of every switching cycle (i.e., the initial condition of each cycle is the same), the outputs can be fully decoupled from one another in the time domain, thereby allowing them to be operated independently. What this means is that I_L can theoretically be

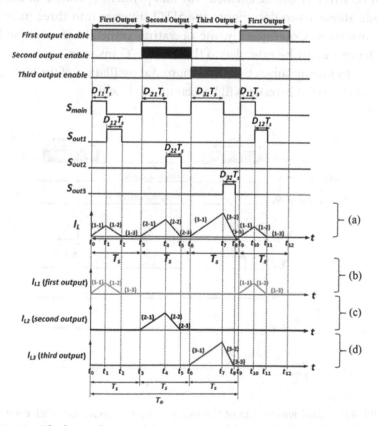

FIGURE 4.8 Ideal waveforms of the output enable signals, gate-drive signals of the switches, and inductor current in the SITO inverter (with all outputs enabled).

decomposed into three separate currents, i.e., I_{L1}, I_{L2}, and I_{L3}, where I_{L1}, I_{L2}, and I_{L3} are the "hypothetical" inductor current for the first output, second output and third output, respectively. Such decomposition of I_L is graphically illustrated in Figure 4.8. The ideal waveform of I_L is annotated as (a), while those of I_{L1}, I_{L2}, and I_{L3} are annotated as (b), (c), and (d), respectively. Conversely, I_L can be reconstructed by combining I_{L1}, I_{L2}, and I_{L3} using the property of superposition. Assume, without loss of generality, that only the first output is enabled (while the second and third outputs are disabled). Figure 4.9 shows the ideal waveforms of the key signals of the SITO inverter with only the first output enabled. It is interesting to note that the waveform of I_L in Figure 4.9 is the same as that of I_{L1} in Figure 4.8. In other words, the SITO inverter behaves like a SISO inverter since only one output is in operation. Likewise, only the second (or third) output of the SITO inverter can be enabled. The independent operation of the three outputs allows us to decompose the SITO inverter into three individual SISO inverters. Consequently, the operating principle of a simple SISO inverter can easily be extended to that of a SITO inverter.

In the following subsection, the sinusoidal oscillation of the first output phase of the SITO inverter will be examined closely.

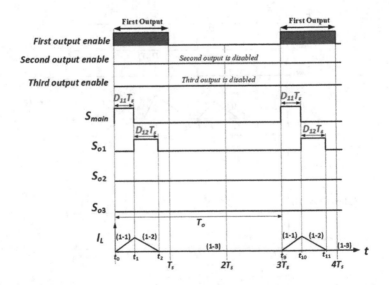

FIGURE 4.9 Ideal waveforms of the output enable signals, gate-drive signals of the switches, and inductor current in the SITO inverter (with only the first output enabled).

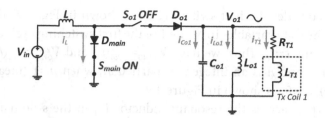

FIGURE 4.10 Equivalent circuit diagram of the first output phase of SITO inverter in Mode 1.

4.5.1 Proof of Sinusoidal Oscillation in Mode 1

The equivalent circuit of the SITO inverter operating in Mode 1 with only the first output enabled is shown in Figure 4.10. Basically, this is a SISO inverter. For simplicity, ideal circuit components with zero losses are assumed. Since the output switch is in the OFF position, the output resonant tank and inductive load are completely separated from the main inductor and the DC power source.

By invoking KCL at the output node, the sum of the three output branch currents can be written as follows.

$$i_{Co1}(t) + i_{Lo1}(t) + i_{T1}(t) = 0 \tag{4.4}$$

By applying Laplace transform to equation (4.4), we have

$$I_{Co1}(s) + I_{Lo1}(s) + I_{T1}(s) = 0 \tag{4.5}$$

Since the resonant inductor (L_{o1}), the resonant capacitor (C_{o1}), and the inductive load (L_{T1}) are connected in parallel, we have $v_{o1}(t) = v_{Co1}(t) = v_{Lo}(t)$. The current across C_{o1} in the time domain can be expressed in the following manner.

$$i_{Co1}(t) = C_{o1}\frac{dv_{Co1}(t)}{dt} = C_{o1}\frac{dv_{o1}(t)}{dt} \tag{4.6}$$

By applying Laplace transform, equation (4.6) can be rewritten as

$$I_{Co1}(s) = C_{o1}\left[sV_{Co1}(s) - V_{Co1,t0}\right] = C_{o1}\left[sV_{o1}(s) - V_{o1,t0}\right] \tag{4.7}$$

where $V_{Co1,t0}$ and $V_{o1,t0}$ are the initial values of the resonant capacitor voltage and the output voltage in Mode 1 of the first switching cycle, respectively.

Let us consider the first switching cycle shown in Fig. 4.9. The initial values of the state variables in Mode 1 of the first switching cycle are determined at time $t=t_0$. Thus, we have $V_{co1,t0}=v_{co1}(t_0)$ and $V_{o1,t0}=v_{o1}(t_0)$, where $t_0=(n-1)T_o$, and n is an integer. In particular, when $n=1$ (meaning the first cycle), $t_0=0$, as shown in Figure 4.9.

The current across the resonant inductor L_{o1} in the s-domain can be written as

$$I_{Lo1}(s)=\frac{V_{Lo1}(s)}{sL_{o1}}+\frac{I_{Lo1,t0}}{s}=\frac{V_{o1}(s)}{sL_{o1}}+\frac{I_{Lo1,t0}}{s} \tag{4.8}$$

where $I_{Lo1,t0}$ is the initial value of the resonant inductor current in Mode 1, i.e., $I_{Lo1,t0}=i_{Lo}(t_0)$.

Since the AC load consists of the resistance R_T in series with the self-inductance L_T of the transmitting coil, the current across the AC load can therefore be expressed as

$$I_{T1}(s)=\frac{V_{o1}(s)+L_{T1}I_{T1,t0}}{R_{T1}+sL_{T1}} \tag{4.9}$$

where $I_{T1,t0}$ is the initial value of the current across the transmitting coil in Mode 1, i.e., $I_{T1,t0}=I_{T1}(t_0)$.

By substituting equations (4.7)–(4.9) into (4.5), we have

$$C_{o1}\left[sV_{o1}(s)-V_{o1,t0}\right]+\frac{V_{o1}(s)}{sL_{o1}}+\frac{I_{Lo1,t0}}{s}+\frac{V_{o1}(s)+L_{T1}I_{T1,t0}}{R_{T1}+sL_{T1}}=0 \tag{4.10}$$

By arranging the terms in equation (4.10), the output voltage $V_{o1}(s)$ can be written as

$$V_{o1}(s)=C_{o1}V_{o1,t0}\left[\frac{sL_{o1}R_{T1}+s^2L_{o1}L_{T1}}{s^2L_{o1}C_{o1}(R_{T1}+sL_{T1})+s(L_{o1}+L_{T1})+R_{T1}}\right]$$

$$-\frac{I_{Lo1,t0}}{s}\left[\frac{sL_{o1}R_{T1}+s^2L_{o1}L_{T1}}{s^2L_{o1}C_{o1}(R_{T1}+sL_{T1})+s(L_{o1}+L_{T1})+R_{T1}}\right] \tag{4.11}$$

$$-\frac{L_{T1}I_{T1,t0}}{R_{T1}+sL_{T1}}\left[\frac{sL_{o1}R_{T1}+s^2L_{o1}L_{T1}}{s^2L_{o1}C_{o1}(R_{T1}+sL_{T1})+sL_{T1}+R_{T1}}\right]$$

To ensure that the resonant frequency is *not* affected by the load inductance, $L_{T1}\gg L_{o1}$. Hence, equation (4.11) can be simplified as

$$V_{o1}(s) = a_1\left(\frac{s}{s^2 + \omega_o^2}\right) + a_2\left(\frac{1}{s^2 + \omega_o^2}\right) + a_3\left(\frac{1}{s + \dfrac{R_T}{L_T}}\right) \quad (4.12)$$

where

$$a_1 = V_{o1,t0} - I_{T1,t0}\frac{R_{T1}L_{o1}L_{T1}}{L_{T1}^2 + R_{T1}L_{o1}C_{o1}} \quad (4.13a)$$

$$a_2 = -\left[\frac{I_{Lo1,t0}}{C_{o1}} + \frac{I_{T1,t0}}{C_{o1}}\left(\frac{L_{T1}^2}{L_{T1}^2 + R_{T1}^2 L_{o1}C_{o1}}\right)\right] \quad (4.13b)$$

$$a_3 = I_{T1,t0}\left(\frac{R_{T1}L_{o1}L_{T1}}{L_{T1}^2 + R_{T1}^2 L_{o1}C_{o1}}\right) \quad (4.13c)$$

and

$$\omega_o = \frac{1}{\sqrt{L_{o1}C_{o1}}} \quad (4.13d)$$

Now, by applying inverse Laplace transform, equation (4.12) can be converted back to the time domain as follows.

$$v_{o1}(t) = A_1\cos(\omega_o t) + B_1\sin(\omega_o t) + C_1 e^{-\left(\frac{R_{T1}}{L_{T1}}\right)t}$$

$$= A_1\cos(\theta) + B_1\sin(\theta) + C_1 e^{-\left(\frac{R_{T1}}{L_{T1}}\right)t} \quad (4.14)$$

where $A_1 = a_1$, $B_1 = a_2\omega_o^{-1}$, $C_1 = a_3$, and $\theta = \omega_o t$.

Since $R_T \gg L_T$, the value of $e^{-\left(\frac{R_T}{L_T}\right)t}$ becomes zero in steady-state condition. This means that the third term on the R.H.S. of equation (4.14) can be neglected. Therefore, equation (4.14) can be reduced to the following.

$$v_{o1}(t) = A_1\cos(\theta) + B_1\sin(\theta) \quad (4.15)$$

Let $\sin(\alpha)=\dfrac{A_1}{\sqrt{A_1^2+B_1^2}}$ and $\cos(\alpha)=\dfrac{B_1}{\sqrt{A_1^2+B_1^2}}$. Hence, equation (4.15) can be rewritten as

$$v_{o1}(t)=\left(\sqrt{A_1^2+B_1^2}\right)\left[\cos(\theta)\sin(\alpha)+\sin(\theta)\cos(\alpha)\right]$$

$$=\left(\sqrt{A_1^2+B_1^2}\right)\sin(\theta+\alpha) \tag{4.16}$$

$$=\left(\sqrt{A_1^2+B_1^2}\right)\sin(\omega_o t+\alpha)$$

where $t_0 \le t < t_1$ for the first switching cycle. In general, for the $(3n-2)^{\text{th}}$ switching cycle, we have $t_{9(n-1)} \le t < t_{9n-8}$, where $t_{9(n-1)}=3(n-1)T_s$, $t_{9n-8}=[3(n-1)+D_{11}]T_s$, and n is an integer.

Equation (4.16) shows that in Mode 1, the SIMO inverter produces a pure sinusoidal output voltage whose frequency is the same as the resonant frequency ω_o of the LC resonant circuit. Intuitively, the first output switch is turned OFF during Mode 1, which disconnects the main inductor and the DC input voltage from the LC tank circuit. Hence, the tank circuit is self-oscillating at the resonant frequency.

4.5.2 Proof of Sinusoidal Oscillation in Mode 2

The equivalent circuit of the first output phase of the SITO inverter operating in Mode 2 is depicted in Figure 4.11. S_{main} is OFF, while S_{o1} is ON. The energy stored in the main inductor L is transferred to the output. By using KVL, we can write

$$V_{in}-v_L(t)-v_{o1}(t)=0 \tag{4.17}$$

where $v_L(t)$ is the voltage across the main inductor L, and $v_{o1}(t)$ is the AC output voltage.

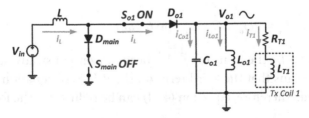

FIGURE 4.11 Equivalent circuit diagram of the first output phase of SITO inverter in Mode 2.

By performing the Laplace transform of equation (4.17), we have

$$\frac{V_{in}}{s} - L\left[sI_L(s) - I_{L,pk1}\right] - V_{o1}(s) = 0 \tag{4.18}$$

where $I_L(s)$ is the inductor current, $I_{L,pk1}$ is the peak value of the inductor current for the first output, and $V_{o1}(s)$ is the first output voltage.

Since $I_{L,pk1} = \dfrac{V_{in}D_{11}T_s}{L}$, equation (4.18) can be reexpressed as

$$\frac{V_{in}}{s} - sLI_L(s) + V_{in}D_{11}T_s - V_{o1}(s) = 0 \tag{4.19}$$

where D_{11} represents the on-time duty ratio in Mode 1.

By referring to the circuit diagram in Figure 4.11 and applying KCL, we have

$$i_L(t) = i_{Co1}(t) + i_{Lo1}(t) + i_{T1}(t) \tag{4.20}$$

The Laplace transform of equation (4.20) can be written as

$$I_L(s) = I_{Co1}(s) + I_{Lo1}(s) + I_{T1}(s) \tag{4.21}$$

By substituting equations (4.7)–(4.9) into (4.21), we have

$$I_L(s) = C_{o1}\left[sV_{o1}(s) - V_{o1,t1}\right] + \frac{V_{o1}(s)}{sL_{o1}} + \frac{I_{Lo1,t1}}{s} + \frac{V_{o1}(s) + L_{T1}I_{T1,t1}}{R_{T1} + sL_{T1}} \tag{4.22}$$

where $V_{o1,t1}$, $I_{Lo1,t1}$, and $I_{T1,t1}$ are the initial values of the output voltage, the resonant inductor current, and the current across the transmitting coil during Mode 2.

Note that the initial values of the state variables in Mode 2 of the first switching cycle are determined at $t = t_1$. In general, for the $(3n-2)^{th}$ switching cycle, the initial values of the state variables in Mode 2 are determined at $t_{9n-8} = [3(n-1) + D_{11}]T_s$, where n is an integer. Mathematically, we have $V_{o1,t1} = v_{o1}(t_1)$, $I_{Lo1,t1} = i_{Lo}(t_1) = I_{L,pk1}$, $I_{T1,t1} = i_{T1}(t_1)$. In particular, when $n=1$ (meaning the first cycle), $t_1 = D_{11}T_s$, as is evident in Figure 4.9.

By substituting equation (4.22) into (4.19) and rearranging, we have

$$V_{o1}(s) = \left[V_{o1,t1} - \frac{R_{T1}L_{o1}LI_{L,pk1}}{L_{T1}\left(L + L_{o1} + \dfrac{R_{T1}{}^2 LL_{o1}C_{o1}}{L_{T1}{}^2} \right)} \right] \left(\frac{s}{s^2 + \dfrac{1}{LC_{o1}} + \dfrac{1}{L_{o1}C_{o1}}} \right)$$

$$- \left(\frac{I_{L,pk1}}{C_{o1}} + \frac{I_{T1,t1}}{C_{o1}} \frac{L + L_{o1}}{L + L_{o1} + \dfrac{R_{T1}{}^2 LL_{o1}C_{o1}}{L_{T1}{}^2}} \right) \left(\frac{1}{s^2 + \dfrac{1}{LC_{o1}} + \dfrac{1}{L_{o1}C_{o1}}} \right)$$

(4.23)

$$+ \left[\frac{R_{T1}L_{o1}LI_{T1,t1}}{L_{T1}\left(L + L_{o1} + \dfrac{R_{T1}{}^2 LL_{o1}C_{o1}}{L_{T1}{}^2} \right)} \right] \left(\frac{1}{s + \dfrac{R_{T1}}{L_{T1}}} \right)$$

$$+ \left(\frac{V_{in}D_{11}T_s + \dfrac{V_{in}}{s}}{LC_{o1}} \right) \left(\frac{1}{s^2 + \dfrac{1}{LC_{o1}} + \dfrac{1}{L_{o1}C_{o1}}} \right)$$

Let $\omega_1 = \sqrt{\dfrac{1}{LC_{o1}} + \dfrac{1}{L_{o1}C_{o1}}}$ and take the inverse Laplace transform of equation (4.23); then, the output voltage in the time domain can be expressed as

$$v_{o1}(t) = A_2 \cos(\varphi) + B_2 \sin(\varphi) + C_2 e^{-\left(\frac{R_{T1}}{L_{T1}} \right)t} + k \qquad (4.24)$$

where

$$A_2 = V_{o1,t1} - \frac{R_{T1}L_{o1}LI_{L,pk1}I_{T1,t1}}{L_{T1}\left(L + L_{o1} + \dfrac{LR_{T1}{}^2 L_{o1}C_{o1}}{L_{T1}{}^2} \right)} - \frac{V_{in}L_{o1}}{L + L_{o1}}, \qquad (4.25a)$$

$$B_2 = -\frac{1}{\omega_1}\left(\frac{I_{L,pk}}{C_{o1}} + \frac{I_{T1,t1}}{C_{o1}} \cdot \frac{L + L_{o1}}{L + L_{o1} + \dfrac{R_T{}^2 LL_{o1}C_{o1}}{L_{T1}{}^2}} - \frac{V_{in}D_{11}T_sL_{o1}}{C_{o1}L} \right), (4.25b)$$

$$C_2 = \frac{I_{T1,t1}R_{T1}L_{ol}L}{L_{T1}\left(L + L_{ol} + \dfrac{R_T^2 LL_{ol}C_{ol}}{L_{T1}^2}\right)}, \text{and} \tag{4.25c}$$

$$k = \frac{V_{in}L_{ol}}{L + L_{ol}}u(t), \tag{4.25d}$$

where $u(t)$ is the unit step function, and $\varphi = \omega_1 t$.

Since $R_T \gg L_T$, the third term on the R.H.S. of equation (4.24) drops out in steady-state condition, and thus, equation (4.24) can be reduced to

$$v_{ol}(t) = A_2\cos(\varphi) + B_2\sin(\varphi) + k \tag{4.26}$$

where k is a fixed DC offset that represents the energy boost from the main inductor and the DC input source.

Suppose $\sin(\beta) = \dfrac{A_2}{\sqrt{A_2^2 + B_2^2}}$ and $\cos(\beta) = \dfrac{B_2}{\sqrt{A_2^2 + B_2^2}}$. Equation (4.26) can be reexpressed as

$$v_{ol}(t) = \left(\sqrt{A_2^2 + B_2^2}\right)\left[\cos(\varphi)\sin(\beta) + \sin(\varphi)\cos(\beta)\right] + k$$

$$= \left(\sqrt{A_2^2 + B_2^2}\right)\sin(\varphi + \beta) + k \tag{4.27}$$

$$= \left(\sqrt{A_2^2 + B_2^2}\right)\sin(\omega_1 t + \beta) + k$$

Basically, equation (4.27) shows that when the inverter operates in Mode 2, the output voltage $v_{ol}(t)$ is a pure sinusoidal signal whose radian frequency is given by $\omega_1 = \sqrt{\dfrac{1}{LC_{ol}} + \dfrac{1}{L_{ol}C_{ol}}}$. It is worth noting that if the value of the main inductor L was much greater than that of the resonant inductor L_{ol} (i.e., $L \gg L_{ol}$), the radian frequency of the output voltage in Mode 2 would be almost identical to that in Mode 1 or Mode 3 (i.e., $\omega_1 \approx \omega_o$).

4.5.3 Proof of Sinusoidal Oscillation in Mode 3

The equivalent circuit model of the first output phase of the SITO inverter operating in Mode 3 is depicted in Figure 4.12.

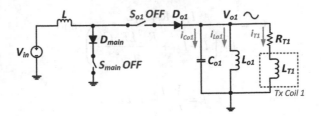

FIGURE 4.12 Equivalent circuit diagram of the first output phase of SITO inverter in Mode 3.

Because the output switch S_{o1} is switched OFF in Mode 3, the LC resonant tank and the inductive load are completely isolated from the main inductor L and the input voltage. As you may recall, S_{o1} is also switched OFF in Mode 1 [refer to Figure 4.10]. Hence, in either Mode 1 or 3, the LC tank circuit self-oscillates with no energy being injected from the main inductor. The frequency of the sinusoidal output voltage in Mode 3 is identical to that of Mode 1, i.e., $\omega_o = \sqrt{\dfrac{1}{L_{o1}C_{o1}}}$. Similar to equation (4.15), the output voltage in Mode 3 can be expressed as

$$v_{o1}(t) = A_3 \cos(\theta) + B_3 \sin(\theta) \tag{4.28}$$

where

$$A_3 = V_{o1,t2} - I_{T1,t2} \frac{R_{T1}L_{o1}L_{T1}}{L_{T1}^2 + R_{T1}L_{o1}C_{o1}}, \tag{4.29a}$$

$$B_3 = -\frac{1}{\omega_0}\left[\frac{I_{Lo1,t2}}{C_{o1}} + \frac{I_{T1,t2}}{C_{o1}}\frac{L_{T1}^2}{L_{T1}^2 + R_{T1}^2 L_{o1}C_{o1}}\right], \text{and} \tag{4.29b}$$

$$\theta = \omega_o t. \tag{4.29c}$$

Note that the initial condition of Mode 3 for the first switching cycle is evaluated at time $t=t_2$. In general, the initial condition of Mode 3 for the $(3n-2)^{\text{th}}$ switching cycle is evaluated at $t_{9n-7}=[3(n-1)+D_{11}+D_{12}]T_s$, where n is an integer. Hence, we have $V_{o1,t2}=v_{o1}(t_2)$, $I_{T1,t2}=i_{T1}(t_2)$, and $I_{Lo1.t2}=i_{Lo1}(t_2)$. In particular, when $n=1$ (meaning the first cycle), $t_2=(D_{11}+D_{12})T_s$, as is evident in Figure 4.9.

Let $\sin(\gamma) = \dfrac{A_3}{\sqrt{A_3{}^2 + B_3{}^2}}$ and $\cos(\gamma) = \dfrac{B_3}{\sqrt{A_3{}^2 + B_3{}^2}}$. Equation (4.28) can be rewritten as

$$v_{o1}(t) = \left(\sqrt{A_3{}^2 + B_3{}^2}\right)\left[\cos(\varphi)\sin(\gamma) + \sin(\varphi)\cos(\gamma)\right]$$

$$= \left(\sqrt{A_3{}^2 + B_3{}^2}\right)\sin(\varphi + \gamma) \qquad\qquad (4.30)$$

$$= \left(\sqrt{A_3{}^2 + B_3{}^2}\right)\sin(\omega_0 t + \gamma)$$

Equation (4.30) shows that the output voltage in Mode 3 resembles a pure sinusoidal signal whose frequency is identical to the resonant frequency ω_o of the LC tank circuit. As far as the first output is concerned, the end of Mode 3 (e.g., $t = t_3$ as shown in Figure 4.9) also marks the end of the present (resonant) cycle and the beginning of the next cycle.

4.6 SIMULATION VERIFICATION

To verify the functionality of the SITO boost inverter, simulations are conducted using PSIM software. Table 4.1 shows the design specifications of the SITO inverter. Without loss of generality, it is assumed that each output is connected to a pure resistive load.

4.6.1 SISO Inverter

The functionality of the SISO inverter is verified by simulations using PSIM software. A SISO inverter can be easily obtained from the SITO inverter

TABLE 4.1 Design Specifications of the SITO Boost Inverter

Design Parameter	Value
Input voltage (V_{in})	3.7 V
Switching frequency (f_{sw})	111 kHz
Main inductor (L)	2.4 µH
ESR of the main inductor (measured at 333 kHz)	20 mΩ
Capacitor in the resonant tank (C_{o1}, C_{o2}, C_{o3})	0.3 µF
ESR of the resonant capacitor (measured at 111 kHz)	6 mΩ
Inductor in the resonant tank (L_{o1}, L_{o2}, L_{o3})	6.8 µH
ESR of the resonant inductor (measured at 111 kHz)	15 mΩ
Load resistor (R_{T1}, R_{T2}, R_{T3})	50 Ω
Forward voltage drop (V_F) across the diode	0.25 V

by enabling only the first output while disabling the other two outputs. The on-time duty ratio of the main switch is initially chosen to be 10%. Since only one output is enabled, the main switch and the first output switch operate at the same frequency, i.e., 111 kHz. In other words, the switching frequency is the same as the output resonant frequency. Figure 4.13 shows the simulated waveforms of the key signals such as the inductor current (I_L), the gate-drive signal of the main switch (S_{main}), the gate-drive signal of the output switch (S_{o1}), and the AC output voltage (V_{o1}). The simulation result shows that the SISO inverter operates in DCM as I_L returns to zero in every switching cycle. It also produces a sinusoidal-like output voltage whose RMS value is around 4.61 V and fundamental frequency is 111 kHz.

Now, the on-time duty ratio of S_{main} is increased from 10% to 20%, while the values of other design parameters remain unchanged. It is envisaged that the RMS value of the output voltage will increase as more energy is stored in the main inductor during the charging phase and is then delivered to the output load during the discharging phase. Figure 4.14 shows the simulated waveforms of the key signals of the SISO inverter.

As shown in Figure 4.14, the SISO inverter continues to operate in DCM. The RMS value of the sinusoidal-like output voltage is increased to 7.64 V. Thus, it can be seen that by increasing the on-time duty ratio of S_{main} (i.e., the charging phase of the main inductor is increased), a higher RMS value of the output voltage can be obtained at a given resistive load. The fundamental frequency of the output voltage remains unchanged at 111 kHz.

4.6.2 SITO Inverter

First, the SITO inverter operating under balanced load condition will be examined. By definition, the term "balanced load" refers to a specific condition in which the *same* energy is delivered across all outputs. This can be achieved by assigning the same on-time duty ratios of S_{main} across the three outputs. Since the load resistance of each output of the SITO inverter is assumed to be the same, the RMS values of the three output voltages should be equal. Figure 4.15 shows the simulated waveforms of the key signals of the SITO boost inverter with balanced loads. Such key signals include the inductor current (I_L), the three output voltages (V_{o1}, V_{o2}, V_{o3}) as well as the gate-drive signals of the main switch (S_{main}) and the output switches (S_{o1}, S_{o2}, S_{o3}).

As expected, the simulation result shows that the three output voltages have the same RMS values of 4.61 V and the same fundamental frequency of 111 kHz. Since this is a boost inverter, the RMS value of the

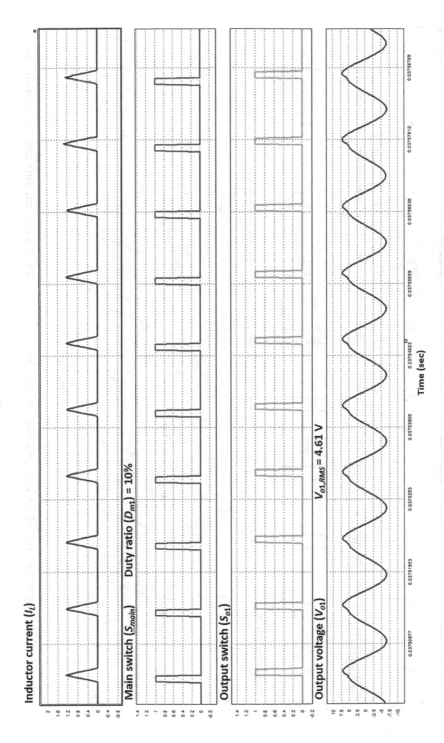

FIGURE 4.13   Simulated waveforms of the key signals of the SISO boost inverter with 10% on-time duty ratio of the main switch.

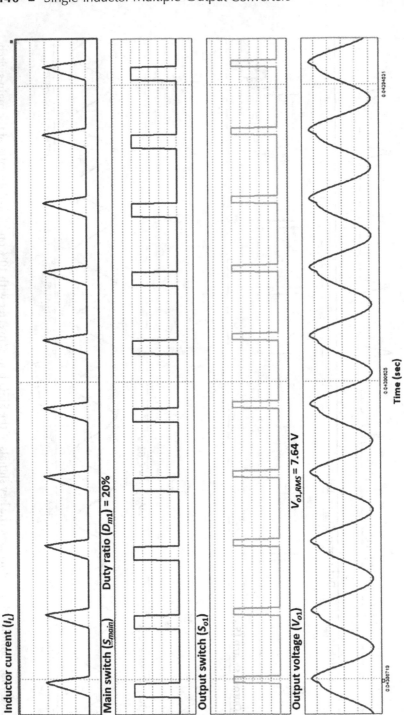

FIGURE 4.14 Simulated waveforms of the key signals of the SISO boost inverter with 20% on-time duty ratio of the main switch.

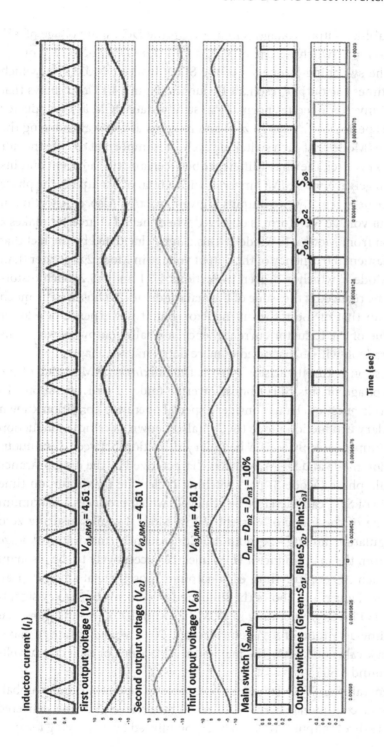

FIGURE 4.15   Simulated waveforms of the key signals of the SITO boost inverter under balanced load condition.

sinusoidal-like output voltage is higher than the DC input voltage of 3 V. In addition, the switching rate of S_{main} is *three times* that of S_{o1}, S_{o2}, or S_{o3}. Hence, the switching frequency of the SITO inverter is 333 kHz, which is three times the output resonant frequency. Figure 4.15 also shows that the SITO inverter operates normally in steady-state DCM as the inductor current is periodically reset to zero and remains at zero value during the idle phase (Mode 3) of each switching cycle. It is important to note that for this SITO inverter, the phase difference between any two adjacent outputs is 120°. In general, for a SIMO inverter with a total of N outputs, the phase difference between any two outputs is given by $2\pi/N$. Also, a distortion in the output voltage waveform is observed when the SITO inverter makes a transition from Mode 1 to Mode 2. This can be elucidated by the fact that the fundamental frequency of the output voltage in Mode 2 is greater than that in Mode 1, which is proven in Sections 4.5.1 and 4.5.2. Such distortion can be reduced if the value of the main inductor is chosen to be much larger than that of the resonant inductor, but at the expense of a lower peak value of the inductor current (hence, a smaller output power). This design tradeoff will be discussed in more detail in later chapters.

A question naturally arises: What is the *maximum* RMS value of the output voltage of this SITO boost inverter under DCM operation? To address this question, let us consider a special operating condition known as boundary conduction mode (BCM) (also known as critical conduction mode). Simply said, BCM is a boundary condition between continuous conduction mode and DCM, which is characterized by the disappearance of the idle phase (Mode 3) per switching cycle. The maximum on-time duty ratio of S_{main} can therefore be achieved, which produces a maximum peak value of the inductor current (I_L) in Mode 1 while ensuring zero cross-regulation among the outputs. Consequently, maximum storage energy from the main inductor (L) is injected successively to the resonant tank at each AC output. Thus, each output can oscillate at a higher peak amplitude. Figure 4.16 shows the simulated waveforms of the key signals of the SITO inverter operating in BCM. The simulation results show that the on-time duty ratio of S_{main} is increased to 70%, while I_L attains a maximum peak value of 2.94 A. The maximum RMS value of each output voltage is around 8.605 V.

In general, the SITO converter is capable of operating under unbalanced load condition whereby different energy levels can be delivered across the three outputs. This can be accomplished by employing *unequal*

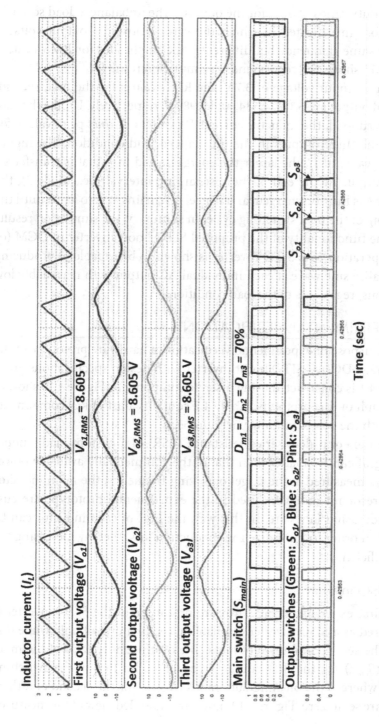

FIGURE 4.16   Simulated waveforms of the key signals of the SITO boost inverter in boundary conduction mode.

on-time duty ratios of S_{main} for the outputs. The unbalanced load scenario is therefore investigated in which the three sinusoidal output voltages have the same frequency but different peak amplitudes (or RMS values). Figure 4.17 shows the corresponding simulation results.

Given an input voltage of 3.7 V, the RMS values for the first, second, and third output are 3.177, 4.534, and 5.989 V, respectively. Unlike the case of balanced load, the inductor current I_L exhibits *distinct* peak values for each of the three outputs in the unbalanced load condition. The higher the peak value of I_L, the larger the energy stored in L that is transferred to the output. At the end of the discharging interval (i.e., Mode 2), the inductor L is fully discharged, and hence, I_L remains at zero value until the beginning of the next switching cycle. In summary, the simulation results verify the functionality of the proposed SITO boost inverter in DCM (or BCM) operation. The SITO inverter is shown to be capable of producing high-quality sine wave at each individual AC output with reasonably low distortions, regardless of the load condition.

4.7 EXPERIMENTAL VERIFICATION

For the purpose of experimental verification, a hardware prototype of the single-stage DC-AC SITO boost inverter with design specifications given in Table 4.1 is constructed using discrete components. Figure 4.18 shows a photograph of this prototype. Table 4.2 contains a list of key components along with the corresponding part numbers.

Since the equivalent series resistance (ESR) of a custom-made inductor is significantly smaller than that of the commercially available power inductors measured at the target operating frequency, the main inductor and all resonant inductors used in the experimental prototype are custom-made using Litz wire. In this way, the ESR of each inductor can be reduced in order to minimize the conduction loss, thereby increasing the power efficiency.

4.7.1 SISO Inverter

In the first experiment, the hardware prototype of the SITO inverter is configured as a SISO inverter by enabling the first output only while disabling the second and third outputs. The on-time duty ratio of the main switch (S_{main}) for the first output is chosen to be 10% of the switching period, whereas the on-time duty ratios of S_{main} for the second and third output are set to zero. Figure 4.19 shows an expanded view of the measured

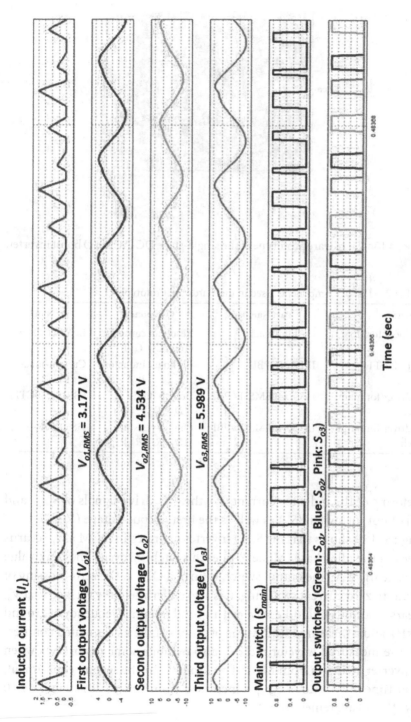

FIGURE 4.17 Simulated waveforms of the key signals of the SITO boost inverter under unbalanced load condition.

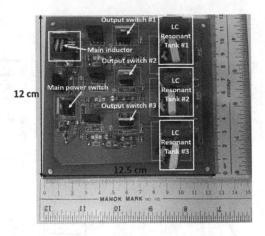

FIGURE 4.18   Hardware prototype of the single-stage DC-AC SITO boost inverter.

TABLE 4.2   List of Components Used in the Hardware Prototype

Component	Part Number	Component	Part Number
Main inductor (L)	Custom-made	Resonant capacitor (C_{o1}, C_{o2}, C_{o3})	ECW-F6304HL
Power MOSFET	IRFZ24NPBF	Resonant inductor (L_{o1}, L_{o2}, L_{o3})	Custom-made
Gate driver for main switch	LTC4440EMS8E#PBF	Branch diode (D_{o1}, D_{o2}, D_{o3})	SBR10U40CTFP
Gate driver for output switch	LTC4440EMS8E#PBF	Output resistor (R_{L1}, R_{L2}, R_{L3})	HS50-50R-F

waveforms of the inductor current (I_L), the gate drive signals of S_{main}, and the first output switch (S_{o1}) as well as the first output voltage (V_{o1}).

Figure 4.19 shows that the SISO inverter operates in DCM as I_L returns to zero in the idle phase (Mode 3) of each switching cycle. The peak value of I_L is nearly 2 A. The output switch is turned OFF at the zero-crossing of I_L to attain zero current switching. It is also experimentally verified that V_{o1} appears as a sinusoidal-like wave with a fundamental frequency of around 111 kHz and a measured RMS value of 5.129 V. Like its simulated counterpart, the measured waveform of V_{o1} also exhibits some distortions when the inverter switches from Mode 1 to Mode 2. In the second experiment, the on-time duty ratio of S_{main} is increased from 10% to 20%. Figure 4.20 shows the measurement results.

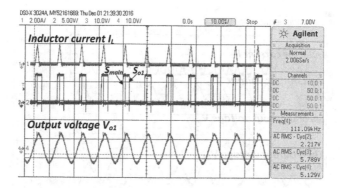

FIGURE 4.19 Measured waveforms of the inductor current (I_L), the gate drive signals of the main switch (S_{main}) and the output switch (S_{o1}), and the first output voltage (V_{o1}) with an on-time duty ratio of 10%.

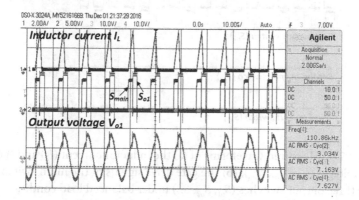

FIGURE 4.20 Measured waveforms of the inductor current (I_L), the gate drive signals of the main switch (S_{main}) and the output switch (S_{o1}), and the first output voltage (V_{o1}) with an on-time duty ratio of 20%.

It can be seen that the peak value of I_L increases to almost 4 A when the duty ratio of S_{main} is doubled. V_{o1} stays fairly sinusoidal with a fundamental frequency of 111 kHz and a measured RMS value of 7.627 V.

4.7.2 SITO Inverter

In the third experiment, the same hardware prototype shown in Figure 4.18 is reconfigured as a SITO boost inverter by enabling all three outputs. Each output is initially connected to a resistive load of the same value. The experiment is then conducted under balanced load condition. Figure 4.21a shows the measured waveforms of the inductor current (I_L) and the three

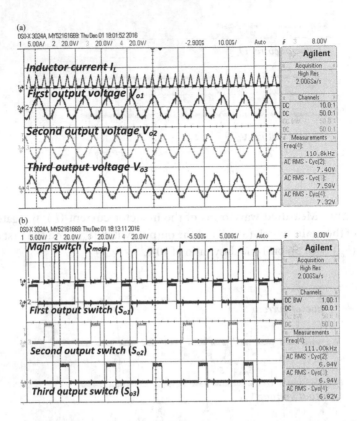

FIGURE 4.21 Measured waveforms of (a) the inductor current (I_L) and the three output voltages (V_{o1}, V_{o2}, V_{o3}); (b) the gate drive signals of the main switch (S_{main}) and the three output switches (S_{o1}, S_{o2}, S_{o3}) under balanced load condition.

output voltages (V_{o1}, V_{o2}, V_{o3}). Figure 4.21b shows the gate drive signals of the main switch (S_{main}) and the three output switches (S_{o1}, S_{o2}, S_{o3}). A 10% on-time duty ratio of S_{main} is used across the three outputs.

The experimental results show that the SITO inverter simultaneously produces three stable sinusoidal-like output voltages with the same frequency of 111 kHz and a phase difference of 120°. The measured RMS values of the three output voltages are 7.40, 7.59, and 7.32 V, respectively. The switching frequency of S_{main} is 333 kHz, which is three times that of the output (resonant) frequency. Now, in the fourth experiment, the SITO boost inverter is used to generate three sinusoidal-like output voltages with *different* amplitudes. In other words, the SITO inverter operates in unbalanced load condition. This is made possible by assigning distinct on-time duty ratios of S_{main} across the three outputs. Figure 4.22a depicts the

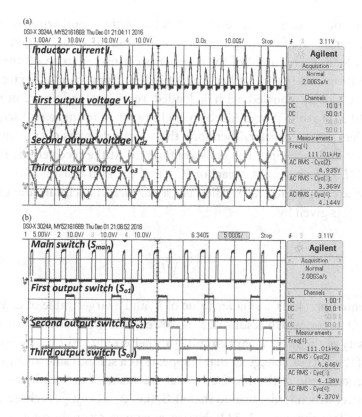

FIGURE 4.22 Measured waveforms of (a) the inductor current (I_l) and the three output voltages (V_{o1}, V_{o2}, V_{o3}); (b) the gate drive signals of the main switch (S_{main}) and the three output switches (S_{o1}, S_{o2}, S_{o3}) under unbalanced load condition.

measured waveforms of I_L, V_{o1}, V_{o2}, and V_{o3}, while Figure 4.22b depicts the gate drive signals of S_{main}, S_{o1}, S_{o2}, and S_{o3}.

It is observed from Figure 4.22a that the inductor current has three distinct peak values corresponding to the three outputs due to the use of unequal on-time duty ratios of S_{main}. The measured RMS values of the first, second, and third outputs are 4.935, 3.369, and 4.144 V, respectively. In this experiment, it becomes evident that the peak (or RMS) value of a particular output voltage is dependent upon the corresponding on-time duty ratio of S_{main}. In other words, a higher on-time duty ratio will result in a larger peak (or RMS) value of the output voltage and vice versa, assuming the other parameters remain constant. The frequency of the three output voltages is around 111 kHz. It is experimentally verified that the SITO inverter can operate properly in DCM under unbalanced load condition.

In practice, the SITO inverter can act as a three-coil wireless power transmitter. In the following experiment, instead of connecting to a pure resistive load, each output of the inverter is connected to an off-the-shelf transmitting coil (part number: 760308100110 from Wurth Electronics), which is closely coupled with a loaded receiving coil (part number: 760308103211 from Wurth Electronics) in order to mimic a real scenario of wireless power transfer. Therefore, three separate pairs of transmitting (Tx) coils and receiving (Rx) coils are formed. At the receiver, the receiving coil (L_o), the resonant capacitor (C_o), and the load resistor (R_l) in each channel are connected in parallel. For a *parallel* RLC circuit, the quality factor (Q) is given by

$$Q = R_L \sqrt{\frac{C_o}{L_o}} \tag{4.31}$$

From equation (4.31), a smaller value of L_o and a larger value of C_o yield a higher Q factor, which results in a narrower bandwidth and increases the selectivity of the LC resonant circuit. This helps reduce the harmonic contents of the output voltage so that it appears as a smooth sinusoidal signal. As a general rule of thumb, the typical reactance of L_o (i.e., $X_L = \omega_o L_o$) at resonance, which is identical to that of C_o (i.e., $X_c = 1/\omega_o C_o$), is chosen to be around a few ohms (e.g. 2–5 Ω). Hence, at a given resonant frequency, the proper values of L_o and C_o can be determined. A more in-depth discussion of the design tradeoff between the total harmonic distortion (THD) and power efficiency will be provided in later chapters.

Figure 4.23 shows the photograph of the experimental setup with three separate pairs of transmitting (Tx) and receiving (Rx) coils. The original SITO inverter is essentially transformed into a low-power three-channel wireless transmitter, which can achieve either simultaneous charging of three independent electronic devices or charging a single three-coil electronic device with greater freedom of positioning.

Figure 4.24a–c shows the frequency spectrum of the sinusoidal output voltage pertaining to the first, second, or third channel at the receiver side under balanced load condition. The measured waveforms of the three output voltages are also shown. The frequency spectrum of a sine wave can be obtained easily by using the fast Fourier transform (FFT) function on an oscilloscope.

As clearly shown in Figure 4.24, the output voltage in each channel is a smooth and clean sine wave with a fundamental frequency of around

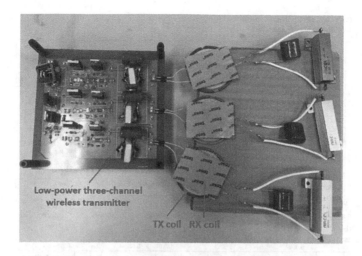

FIGURE 4.23 A photograph of the low-power three-channel wireless power transmitter with three pairs of transmitting and receiving coils.

111 kHz. No noticeable voltage spikes and/or distortions are observed in the measured output voltage waveforms. In addition, the measured peak-to-peak amplitudes for the three output voltages are in close agreement with each other. This comes as no surprise under balanced load condition. Table 4.3 tabulates the measured RMS amplitude of the first harmonic (i.e., fundamental frequency) and the first six harmonics of the fundamental of the first output voltage. The measured THD is only 2.46%, which is sufficiently small.

Last but not least, the unbalanced load condition is also investigated using the same experimental setup shown in Figure 4.23. Figure 4.25 shows the measured waveforms for the three output voltages with *different* peak-to-peak amplitudes. The measured peak-to-peak amplitudes of the first, second, and third output voltages at the receiver are 15.9, 12.1, and 9.6 V. Each output voltage waveform continues to appear as a smooth sine wave with a fundamental frequency of around 111 kHz.

In conclusion, the experimental results demonstrate the effectiveness of the SITO boost inverter in generating three independent sinusoidal output voltages with either identical or different peak values from a single DC power supply. Given an input power of 7.1 W, the measured power efficiency of the SITO inverter at the rated output power of 2 W per channel (i.e., a total output power of 6 W) is around 84.5%.

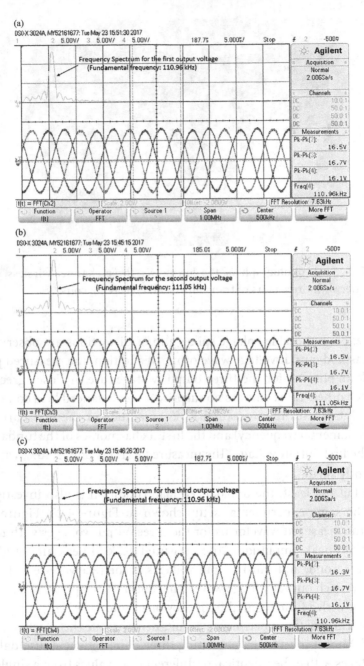

FIGURE 4.24 Frequency spectrum of the sinusoidal output voltage for the (a) first output, (b) second output, and (c) third channel, respectively, measured at the receiver side (as indicated in Channel 4) along with the measured waveforms of the three output voltages.

TABLE 4.3   Measured Harmonic Contents of the First Output Voltage

Frequency (kHz)	Harmonic (#)	Measured V_{rms} (V)
111	1	5.19
222	2	0.126
333	3	0.0148
444	4	0.0094
555	5	0.00403
666	6	0.00403
777	7	0.00574

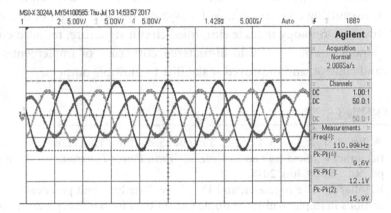

FIGURE 4.25   Measured waveforms for the three sinusoidal output voltages in unbalanced load condition.

4.8 SUMMARY OF KEY POINTS

1. The functions of a DC-DC SIMO converter and a parallel network of LC resonant tanks (acting as the DC-AC stage) are effectively combined to form a new single-stage DC-AC SIMO inverter.

2. A boost-type DC-AC SIMO inverter is introduced, which employs a single inductor in the power stage to produce multiple independent sinusoidal-like output voltages. The main inductor converts the input voltage source into periodic DC current pulses, which are injected into each of the resonant tanks in a round-robin time-multiplexed manner. Subsequently, the resonant tank converts the DC current pulses into an AC power source oscillating at the target resonant frequency.

3. Since this particular type of SIMO inverter is essentially a step-up (boost) converter, the RMS value of each AC output voltage is generally higher than the DC input voltage.

4. The RMS value of the AC output voltage (or output power) can be varied by adjusting the on-time duty ratio of the main switch corresponding to that output. By assigning the same or different on-time duty ratios across the outputs, the RMS value (or amplitude) of each AC output can be independently controlled.

5. A practical application of this SIMO inverter is to drive an array of transmitting coils in a MC-WPT system. The advantages of this new inverter topology include simplified circuit structure, reduced component count, low build-of-material cost, ease of implementation and control, and improved scalability to multiple outputs.

REFERENCES

1. W. Jin, A. T. L. Lee, S. Li, S.-C. Tan, and S. Y. (Ron) Hui, "Low-power multichannel wireless transmitter", *IEEE Trans. Power Electron.*, vol. 33, no. 6, pp. 5016–5028, Jun. 2018.
2. R. Johari, J. V. Krogmeier, and D. J. Love, "Analysis and practical considerations in implementing multiple transmitters for wireless power transfer via coupled magnetic resonance", *IEEE Trans. Ind. Electron.*, vol. 61, no. 4, pp. 1774–1783, Apr. 2014.
3. M. Q. Nguyen, Y. Chou, D. Plesa, S. Rao, and J.-C. Chiao, "Multiple-inputs and multiple- outputs wireless power combining and delivering systems", *IEEE Trans. Power Electron.*, vol. 30, no. 11, pp. 6254–6263, Nov. 2015.
4. B. H. Waters, B. J. Mahoney, V. Ranganathan, and J. R. Smith, "Power delivery and leakage field control using an adaptive phase array wireless power system", *IEEE Trans. Power Electron.*, vol. 30, no. 11, pp. 6298–6309, Nov. 2015.
5. L. Shi, Z. Kabelac, D. Katabi, and D. Perreault, "Wireless power hotspot that charges all of your devices", *Proceedings of 2015 Annual International Conference on Mobile Computing*, pp. 2–13, Sep. 2015.
6. Application Note: NXP Semiconductors, "Coils used for wireless charging," Eindhoven, The Netherlands (2014). [Online] Available: http://cache.freescale.com/files/microcontrollers/doc/app_note/AN4866.pdf.
7. J. Lee, et al., "Wireless power transmitter and wireless power transfer method thereof in many-to-one communication," U.S. Patent US9306401 B2, 2011.
8. A. H. Mohammadian, et al., "Wireless power transfer using multiple transmit antennas," U.S. Patent US8629650 B2, 2008.

9. Datasheet: MWCT1200DS, "MWCT1200DS," NXP Semiconductors (2015). [Online] Available: http://cache.nxp.com/files/microcontrollers/doc/data_sheet/MWCT1200DS.pdf?fpsp=1&WT_TYPE=Data%20Sheets&WT_VENDOR=FREESCALE&WT_FILE_FORMAT=pdf&WT_ASSET=Documentation&fileExt=.pdf.

10. User's Guide: WCT1001A/WCT1003A, "WCT1001A/WCT1003A automotive A13 wireless charging application user's guide," NXP Semiconductors (2014). [Online] Available: http://cache.nxp.com/files/microcontrollers/doc/user_guide/WCT100XAWCAUG.pdf.

11. J. Burdio, F. Monterde, J. Garcia, L. Barragan, and A. Martinez, "A two-output series- resonant inverter for induction-heating cooking appliances", *IEEE Trans. Power Electron.*, vol. 20, no. 4, pp. 815–822, Jul. 2005.

PROBLEMS

4.1 State the advantages of using the SIMO boost inverter topology to implement an inverter with multiple outputs, as compared to using multiple full-bridge inverters?

4.2 For a SIMO boost inverter loaded with inductive coils, describe the necessary condition for attaining smooth sinusoidal output voltages with minimum distortion. Explain why this condition is not always easy to satisfy.

4.3 For an SITO boost inverter, the value of the main inductor is chosen to be 2.7 μH. The values of the resonant inductor and resonant capacitor are 6.8 μH and 0.37 μF, respectively. Determine the switching frequency of the SITO boost inverter.

4.4 For the SITO boost inverter of Problem 4.3, the input voltage is 3.7 V. Each of the three outputs of the SITO boost inverter is loaded with a 47 Ω resistor. Ignore the parasitic resistance of the passive components. Create a circuit (simulation) model of the SITO boost inverter. Simulate this circuit using one of the simulation software (e.g. Pspice, PSIM, Simulink, etc.). Plot the waveforms of the inductor current and the three output voltages when this inverter operates in DCM.

4.5 Based on your simulation model of the SITO boost inverter in Problem 4.4, suggest a way to increase the total output power without making any circuit modification. Determine the maximum total output power that can be achieved by this SITO boost inverter without cross-regulation.

4.6 A SIDO boost resonant inverter can be configured as a two-channel wireless power transmitter (wireless charger) by connecting an off-the-shelf transmitting coil to each AC output. The transmitting coil in each channel is tightly coupled with the corresponding loaded receiving coil for wireless power transfer. The conventional parallel-to-parallel (PP) compensation topology is employed. The mutual inductance between the transmitting (primary) coil and the receiving (secondary) coil is assumed to be constant due to simple fixed positioning charging. On the primary side, the equivalent inductance (L_{eq1}, L_{eq2}) is 24 µH and the equivalent resistance (R_{eq1}, R_{eq2}) is 22 Ω per channel. The resonant frequency is 100 kHz. Figure 4.26 shows the ideal SIDO inverter with the equivalent parallel imped-ance on the primary side. The following is a summary of the design parameters:

$$V_{in} = 3.7\,\text{V}$$
$$L_1 = 3\,\mu\text{H}$$
$$L_{eq1} = L_{eq2} = 24\,\mu\text{H}$$
$$R_{eq1} = R_{eq2} = 22\,\Omega$$
$$C_{o1} = C_{o2} = 0.37\,\mu\text{F}$$
$$f_{o1} = f_{o2} = 100\,\text{kHz}$$

For simplicity, all losses can be neglected.

a. Determine the value of the resonant inductor L_{o1} (or L_{o2}). Round it to one decimal place.

b. The on-time duty ratio of the main power MOSFET (S_m) is set to 60%. Calculate the peak value of the inductor current.

c. The SIDO boost inverter operates in DCM when the on-time duty ratio of the main switch (S_{main}) is 60%. Estimate the peak value of the output voltage at the resonant frequency.

4.7 In the SIDO boost inverter of Figure 4.26, three blocking diodes are used to prevent unintended reverse current flow. Yet the forward voltage drop across these diodes leads to conduction losses, which lowers the overall power efficiency. For the purpose of reducing the

conduction losses, modify the original circuit in Figure 4.26 by eliminating the blocking diodes while still achieving the same circuit
functionality. Draw the new circuit with your proposed changes.

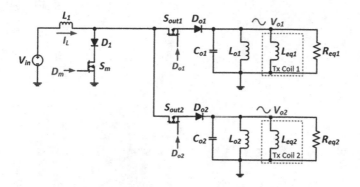

FIGURE 4.26   Single-inductor dual-output (SIDO) boost inverter with the parallel equivalent circuit.

CHAPTER 5

Single-Inductor Multiple-Output DC-AC Buck-Boost Inverters

5.1 INTRODUCTION

Similar to its DC-DC counterpart, the single-inductor multiple-output (SIMO) inverter can be configured as a buck-only, boost-only, or buck-boost type. In this chapter, a new variant of the SIMO inverter known as the buck-boost SIMO inverter will be introduced. Before delving into the details of this particular inverter, it is important to understand the background and motivation of this work and how this inverter can benefit multi-coil wireless power transfer.

Wireless power transfer (WPT) technology has become increasingly popular in recent years. It enables safe and convenient wireless charging of a myriad of portable electronic devices including smartphones, tablets, laptops, wearables, biomedical implants, etc. In the early days, a two-coil approach is employed to realize WPT such as the inductive charging of a wireless electric toothbrush. It simply consists of a pair of transmitting and receiving coils. Unfortunately, a major drawback of this two-coil system is that only a single device can be charged at a time. It also suffers from a low power transfer efficiency due to coil misalignment and the actual separation between the transmitting and receiving coils [1–4]. To

DOI: 10.1201/9781003239833-5

FIGURE 5.1 A printed circuit board (PCB) of a Qi-compliant four-coil wireless power transmitter. (Courtesy of Convenient Power Limited.)

overcome these shortcomings, a multi-coil approach has quickly emerged in which the wireless power transmitter (or wireless charger) contains multiple coils, typically arranged in an array. Figure 5.1 shows a real four-coil wireless power transmitter which complies with the Qi standard [5,6].

The use of multiple transmitting coils increases the likelihood that one of the transmitting coils on the charging pad will align properly with the receiving coil on the device being charged. This substantially improves the freedom of positioning of the end device, which offers greater convenience for users without compromising the power efficiency. A second advantage of the multi-coil charging system is that it enables simultaneous and independent charging of multiple devices. Even though the idea of using only a single wireless charger to power multiple devices at the same time is immensely appealing, it poses a major challenge to the realization of the multi-coil wireless power transmitter due to increased design complexity and higher cost. Compared with the conventional inverter topologies [7–14], the SIMO-based inverter topology appears to be a much more promising solution, which exhibits a simplified circuit structure, smaller form factor, lower cost, and enhanced flexibility and scalability. Nonetheless, the boost-only single-inductor three-output (SITO) inverter discussed in Chapter 4 suffers from two major drawbacks. First, its maximum output power is rated at a modest 2 W per channel which is hard to keep pace with the rapidly growing energy consumption of smartphones and other mobile devices. Second, the RMS value of the sinusoidal output voltage must always be greater than the DC value of the input voltage. This imposes unnecessary restrictions on the voltage requirement of the end devices for wireless charging.

To address the limitations of its predecessor, a novel buck-boost SIMO inverter is therefore introduced [15]. It aims at providing a single-stage DC-AC power conversion from a DC power source into multiple independent AC output voltages whose RMS values can be greater than or

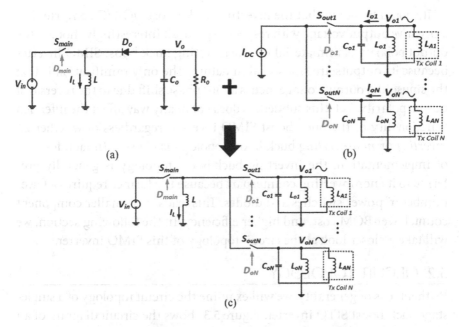

FIGURE 5.2 Derivation of a single-stage buck-boost SIMO inverter. (a) An inverting buck-boost DC-DC converter. (b) A current source driving a parallel network of LC resonant tanks. (c) A single-stage buck-boost DC-AC SIMO inverter.

less than the DC input voltage. Also, by operating the SIMO inverter in pseudo-continuous conduction mode (PCCM), a higher output power with improved efficiency can be achieved. Basically, this new inverter topology is formulated by integrating the functions of an inverting buck-boost DC-DC converter and an independent DC current source driving a parallel network of LC resonant tanks, which acts as the DC-AC stage. Figure 5.2 gives a graphical illustration of the derivation of this topology.

The independent current source (I_{DC}) in Figure 5.2b can be effectively replaced by the main inductor L in Figure 5.2a since the latter acts as a current source that releases its storage energy successively to each of the AC outputs. Due to the fact that the buck-boost converter and the resonant tanks share a common ground, the DC-DC and DC-AC conversion stages can be combined naturally and easily to form a single-stage DC-AC inverter. This results in a buck-boost DC-AC SIMO inverter with a total of N sinusoidal AC outputs, as shown in Figure 5.2c. As the name suggests, only a single inductor in the power stage is needed to drive multiple AC outputs.

It should be noted that the inverting buck-boost DC-DC converter has a *negative* output voltage with respect to ground. Interestingly though, the output polarity is immaterial for the inverting buck-boost SIMO inverter because its outputs are sinusoidal in nature. The only ramification is that the sinusoidal output voltage incurs a 180° phase shift due to the reversal of output polarity. Yet, this subtleness does *not* in any way affect the intended functionality of the buck-boost SIMO inverter, regardless of whether an *inverting* or *noninverting* buck-boost topology is chosen. In fact, for ease of implementation, the inverting buck-boost topology is generally preferred to its noninverting counterpart because the former requires a fewer number of power switches and diodes. This leads to a smaller component count, lower BOM cost, and higher efficiency. In the following section, we will take a closer look at the circuit topology of this SIMO inverter.

5.2 CIRCUIT TOPOLOGY

Without loss of generality, we will examine the circuit topology of a single-stage buck-boost SITO inverter. Figure 5.3 shows the circuit diagram of an ideal buck-boost SITO inverter. For simplicity, all circuit parasitic elements such as the equivalent series resistance (ESR) of a capacitor and the DC resistance (DCR) of an inductor are neglected. The SITO inverter transforms a single DC input voltage (V_{in}) into three independent AC output

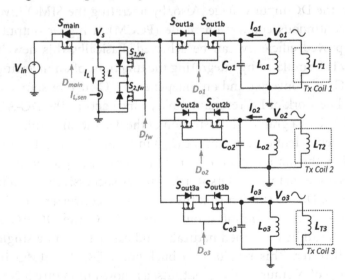

FIGURE 5.3 Ideal circuit diagram of the power stage of the buck-boost SITO inverter.

voltages (V_{o1}, V_{o2}, V_{o3}). The RMS value of each output voltage can be set to be above or below V_{in} by adjusting the corresponding on-time duty ratio of the main switch (S_{main}). The buck-boost SIMO inverter has a total of nine power switches including one main switch (S_{main}), three pairs of output switches [(S_{out1a}, S_{out1b}), (S_{out2a}, S_{out2b}), (S_{out3a}, S_{out3b})], and a pair of freewheeling switches (S_{fw1}, S_{fw2}) across the main inductor L. The current across L is represented by I_L while the output currents across the first, second, and third output branches are represented by I_{o1}, I_{o2}, and I_{o3}, respectively. Each output branch comprises a parallel LC resonant tank (L_{oi}, C_{oi}), which constitutes an integral part of the power stage, along with an inductive load (L_{Ti}) for modeling the transmitting coil, where i is the output index.

By enabling the free-wheeling switches properly, the SITO inverter can operate in PCCM in which the inductor current stays above zero value throughout the idle phase (Mode 3). Due to the nonzero DC offset of the inductor current, the SITO inverter in PCCM is capable of delivering larger output currents than in DCM while still maintaining zero cross-regulation. Hence, a higher output power can be achieved. Rather than using only a single MOSFET, back-to-back connected MOSFETs are employed to break an unintended conduction path via the internal body diode of the MOSFET even when it is in the OFF state. Imagine the top free-wheeling switch ($S_{1,fw}$) is replaced with a short circuit and only the bottom free-wheeling switch ($S_{2,fw}$) stays connected. In the event that S_{out1a} and S_{out1b} are turned ON and the first output voltage (V_{o1}) has a negative value in the second subinterval of PCCM, the switching node V_s will also become negative, which causes an unwanted reverse current to flow from ground to V_s via the body diode of $S_{2,fw}$ (albeit this switch is OFF). Obviously, this is *not* the correct circuit behavior. Hence, two back-to-back free-wheeling MOSFETs are required to avoid a direct short between V_s and ground via the body diode of $S_{2,fw}$. Notice that $S_{1,fw}$ should be oriented in such a way that its internal body diode and that of $S_{2,fw}$ are pointing in the opposite direction.

It should be noted that in the boost-only SITO inverter, a main diode D_1 [see Figure 4.6 in Chapter 4] is used to prevent an unwanted back flow of current from ground to the negative output via the internal body diode of S_{main} during the second subinterval of DCM. In contrast, such a blocking diode is *not* required in the buck-boost SITO inverter, as shown in Figure 5.3. This is because during the second (discharging) subinterval when the output switch is turned ON, the sinusoidal output voltage (and also the voltage at the switching node V_s) has a negative value relative to ground, which is *less than* the input voltage V_{in}. This implies that no reverse current can flow from V_s to V_{in} via the body diode of S_{main}. Hence, unlike its boost-only predecessor, a blocking diode in series with S_{main} is no longer needed in this buck-boost SITO inverter. The elimination of this blocking diode along the main current path means reduced conduction loss, which helps improve the overall power efficiency.

The buck-boost inverter topology also carries another conspicuous advantage over its boost-only predecessor. By taking advantage of the fact that the main inductor L is ground referenced, as shown in Figure 5.3, a simple low-side current-sensing circuitry for the main inductor current I_L can be constructed. For example, the current sensor can be implemented by inserting a small current shunt resistor between L and ground. The voltage across this resistor can be used as a feedback signal which is amplified by a current-sense amplifier for the detection of peak-crossing and valley-crossing of I_L. Because the sense voltage across the current shunt resistor is ground-referenced, a low-side current sensor can be used, which is relatively inexpensive and easy to implement. It allows the current sense amplifier to be a low voltage part as the voltage being sensed is only on the order of ten to a few hundred millivolts above ground. In this way, the sense voltage is not derived from higher voltage, and hence, no common mode rejection is needed.

A pair of back-to-back power MOSFETs (e.g. S_{out1a}, S_{out1b}) is used in each of the three output branches to block an unintended flow of current from V_{in} to the output node during the first subinterval. For illustration purpose, let us consider the first output. Suppose only S_{out1a} remains connected in the first output branch, and S_{out1b} is shorted out. During the first (charging) subinterval, S_{main} is turned ON, while S_{out1a} is turned OFF. When the instantaneous value of V_{o1} becomes smaller than V_{in}, a short circuit current will flow from V_{in} to V_{o1} via the body diode of S_{out1a} even when S_{out1a} is in the OFF state. As a result, a pair of two back-to-back connected power MOSFETs at each output branch is necessary for proper circuit operation. It is a common practice to use power MOSFETs with very low on-resistance ($R_{DS,\,on}$) in order to minimize the forward voltage drop, thereby improving the power efficiency. Their forward voltage drop can be as low as 0.1 V versus the typical Schottky diode forward voltage drop of 0.4–0.5 V. Thus, the conduction loss of a power MOSFET is lower than that of a Schottky diode. Hence, to achieve higher efficiency, the use of back-to-back connected MOSFETs is usually preferred to a MOSFET in series with a blocking diode. In general, the power stage of the buck-boost SIMO inverter requires a total of $(2N+3)$ power MOSFETs in PCCM or $(2N+1)$ power MOSFETs in DCM, where N is the number of outputs.

5.3 OPERATING PRINCIPLE

In this section, the operating principle of a buck-boost SITO inverter is discussed in detail. The same principle can also be extended to a generalized buck-boost SIMO inverter with N outputs. First, the switching sequence of this SITO inverter, accompanied by the corresponding equivalent circuits, is presented.

5.3.1 Switching Sequence and Equivalent Circuits

In the SIMO topology, a general approach is to assign the inductor current (I_L) to each individual output in a round-robin time-multiplexed manner. The use of time-multiplexing control with multiple energizing phases means that the outputs are fully decoupled in time, which avoids undesirable cross-regulation (or cross-interference). By definition, cross-regulation means that a change in one output induces an unwanted change in another output. The elimination of cross-regulation is also a challenge facing the SIMO inverter as well. Figure 5.4a shows the theoretical waveforms of the gate-drive signals of all switches (S_{main}, S_{out1}, S_{out2}, S_{out3}, S_{fw}), the main inductor current (I_L), the input current (I_{input}), and the three AC output voltages (V_{o1}, V_{o2}, V_{o3}). Figure 5.4b shows the corresponding equivalent circuits of each output in three modes of operation. For simplification, the back-to-back MOSFETs are modeled as an ideal switch.

The SIMO inverter operating in PCCM (or DCM) contains three unique modes (or phases) of operation, namely Mode 1 (charging phase), Mode 2 (discharging phase), and Mode 3 (idle phase) per output. For the sake of consistency, the following notations will be used throughout the rest of the chapter. In Figure 5.4a and b, the first number within the parenthesis indicates the output number, while the second number indicates the mode of operation. For instance, (1–1) means the first output and Mode 1. Also, the on-time duty ratios of S_{main} associated with the *first*, *second*, and *third* outputs are uniquely labeled as D_{11}, D_{21}, and D_{31}, respectively. Likewise, the first number in the subscript of the duty ratio denotes the output number, while the second number denotes the mode of operation. The duty ratio of S_{main} is largely determined by the peak limits of I_L. Figure 5.4a depicts the general condition of unbalanced load in which *unequal* maximum

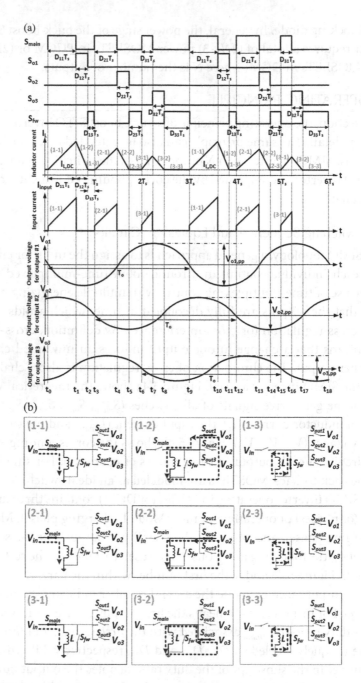

FIGURE 5.4 (a) Theoretical waveforms of the gate drive signals of main switch, output switches, freewheeling switch, inductor current, input current, and three output voltages and (b) the corresponding equivalent circuits of each output in three modes of operation.

(peak) current limits of I_L (i.e., $I_{Lmax1} > I_{Lmax2} > I_{Lmax3}$) are used across the three outputs. Thus, $D_{11} > D_{21} > D_{31}$. Due to the uneven distribution of storage energy from the main inductor to the three outputs, the three output voltages exhibit different peak-to-peak amplitudes. However, if the same peak current limit was used across all outputs, the special condition of balanced load that the three outputs have the same output power level would be achieved. On the other hand, the DC offset of the inductor current ($I_{L, DC}$) shown in Figure 5.4a determines the minimum (valley) limit of I_L. In this case, a constant (positive) value of $I_{L, DC}$ is used, which means that the three outputs share the same valley limit. In particular, for DCM, $I_{L, DC}$ is equal to zero.

5.3.2 Modes of Operation

Figure 5.4 clearly shows that there are three distinct modes of operation per switching period for the buck-boost SITO inverter. Without loss of generality, the modes of operation of this inverter will be elucidated by considering the first output. The same operating principle also applies to the other two outputs.

In Mode 1 (e.g. from time t_0 to t_1), the main switch S_{main} is turned ON, and all other switches are turned OFF. The inductor current I_L ramps up with a rising slope of $m_1 = \dfrac{V_{in}}{L}$. At the end of Mode 1, I_L reaches its peak value ($I_{L, pk}$), which can be mathematically expressed as

$$I_{L,pk} = m_1 D_{11} T_s + I_{L,DC} = \left(\frac{V_{in}}{L} \right) D_{11} T_s + I_{L,DC} \tag{5.1}$$

where D_{11} is the on-time duty ratio of S_{main} for the first output, T_s is the switching period, and $I_{L, DC}$ is the DC offset of the inductor current.

As soon as I_L hits its maximum (peak) current limit (i.e., $I_L = I_{Lmax1}$), a peak-crossing event of I_L occurs, and the inverter switches from Mode 1 to Mode 2. Mode 1 for the first, second, and third output is annotated as (1–1), (2–1), and (3–1), respectively, as shown in Figure 5.4a and b.

In Mode 2 (e.g., from time t_1 to t_2), the main switch S_{main} is turned OFF, and the first output switch S_{out1} is turned ON, while the other two output switches (S_{out2}, S_{out3}) are OFF. The main inductor L releases its storage energy to the first output. The inductor current I_L decreases with a falling slope of $m_2 = \dfrac{V_{o1}(t)}{L}$ until the valley-crossing of I_L (i.e., $I_L = I_{L, DC}$) has occurred. Notice that $V_{o1}(t)$ is the sinusoidal voltage of the first output. In theory, m_2 varies with the instantaneous value of the sinusoidal output voltage. But, as a first-order approximation, $V_{o1}(t)$ can be represented by the average value of the sinusoidal output voltage ($V_{o1,avg}$) during Mode 2. The value of $V_{o1,avg}$ can be obtained as follows.

$$V_{o1,avg} = \frac{1}{D_{12}T_S} \int_{D_1 T_S}^{(D_{11}+D_{12})T_S} V_{o1}(t)dt = \frac{1}{D_{12}T_S} \int_{D_1 T_S}^{(D_{11}+D_{12})T_S} V_m \sin(\omega_0 t + \beta)dt \quad (5.2)$$

where V_m is the amplitude of the sinusoidal output voltage V_{o1}, ω_o is the resonant frequency in radians, and β is the phase angle.

It is crucial to note that for the SITO inverter, the switching period T_s is one-third of the resonant period T_o. Thus, T_s can be expressed as

$$T_S = \frac{T_o}{3} = \frac{2\pi}{3\omega_o} \quad (5.3)$$

In general, for the SIMO inverter, $T_s = \dfrac{T_o}{n}$, where n is the number of AC outputs.

By combining equations (5.2) and (5.3), $V_{o1,avg}$ can be obtained as follows.

$$V_{o1,avg} = \frac{nV_m}{2\pi D_{12}}(\cos\theta_1 - \cos\theta_2) \quad (5.4)$$

where $\theta_1 = \dfrac{2\pi D_{11}}{n} + \beta$ and $\theta_2 = \dfrac{2\pi(D_{11}+D_{12})}{n} + \beta$.

Equation (5.4) can be further reexpressed in terms of the sine function as follows.

$$V_{o1,avg} = \frac{nV_m}{2\pi D_{12}}\left[\sin\left(\frac{\pi}{2}-\theta_1\right) - \sin\left(\frac{\pi}{2}-\theta_2\right) \right] \quad (5.5)$$

Typically, the small-angle approximation is applicable to angles no more than 0.2443 radians (or about 14°), which yields a 1% error. Simply put, the two angles $\left(\dfrac{\pi}{2}-\theta_1\right)$ and $\left(\dfrac{\pi}{2}-\theta_2\right)$ in equation (5.5) must *not* be greater than 0.2443. This imposes an additional design constraint, which is $\dfrac{D_{11}}{n} \geq 0.2111$. Note that D_{11} is always less than 1. Hence, for the SITO inverter (where $n=3$), this design constraint is satisfied whenever we have $0.6333 \leq D_{11} < 1$. As a matter of fact, a higher output power can be obtained by using a higher peak limit of the inductor current, which is accompanied by a larger value of D_{11}. Consequently, this design constraint can be met at higher output power.

By applying the small-angle approximation, equation (5.5) can be reduced to

$$V_{o1,avg} \approx \frac{nV_m}{2\pi D_{12}}(\theta_2 - \theta_1) = V_m \qquad (5.6)$$

Equation (5.6) shows that the average value of the output voltage in Mode 2 can be approximated by the amplitude of the output voltage. Practically speaking, such an approximation is valid as long as the time interval of Mode 2 is much shorter than the resonant period T_o, i.e., $D_{12}T_s \ll T_o$. This implies that either the value of D_{12} is relatively small, or the SIMO inverter has a greater number of outputs. Hence, the average value of the output voltage in Mode 2 becomes virtually constant at a value of V_m. As a result, the downward slope of the inductor current (m_2) can be rewritten as

$$m_2 \approx \frac{V_{o1,avg}}{L} = \frac{V_m}{L} \qquad (5.7)$$

Since the current through an inductor cannot change instantaneously, the direction of the inductor current remains the same between Mode 1 and Mode 2. That is, the inductor current always flows from the switching node to ground via the main inductor. Due to the inverting buck-boost characteristic, it is worth noting that the inductor current actually flows in the *counterclockwise* direction in Mode 2, as shown in Figure 5.4b. Specifically, the current flows from the switching node to ground via the main inductor, and then travels from ground to the output node and eventually returns to the switching node via the output switch. The output switch S_{out1} remains ON until the valley-crossing (or zero-crossing) of I_L is detected in PCCM (or DCM), which indicates the end of Mode 2. Afterwards, S_{out1} is switched OFF, and the inverter makes a transition from Mode 2 to Mode 3. Mode 2 for the first, second, and third output is annotated as (1–2), (2–2), and (3–2), respectively, as shown in Figure 5.4a and b.

In Mode 3 (e.g., from time t_2 to t_3), only the freewheeling switch (S_{fw}) is ON, whereas all other switches (S_{main}, S_{out1}, S_{out2}, S_{out3}) are OFF. This creates a recycling path for the inductor current to "freewheel" between the main inductor and S_{fw}. In principle, the inductor current stays at a constant value of $I_{L,DC}$ in this idle state in PCCM. Yet, in practice, due to the presence of the circuit parasitic elements such as the ESR of the main inductor and the on-resistance of S_{fw}, the main inductor current will not be constant but rather decrease gradually. As shown in Figure 5.4a and b, Mode 3 of the first, second, and third output is annotated as (1–3), (2–3), and (3–3), respectively. In particular, in DCM, the inductor current will return to zero value in Mode 3, i.e., $I_{L,DC}=0$, in which all switches (including S_{fw}) are in their OFF state.

It can be observed that the only time interval that the main inductor L is connected to the output circuit is Mode 2 whereby the storage energy in L is transferred to the output. In Mode 1 and 3, the output circuit is literally self-resonating as it is completely separated from L and the input stage of the inverter. The above mode transition (or switching) process is then repeated for the second output, followed by the third output in the next two switching cycles. Only one output switch can be turned ON at any switching cycle. For example, if the first output is enabled, S_{out1} is ON, whereas S_{out2} and S_{out3} are OFF. This allows the storage energy in L to be distributed across the three outputs sequentially in a time-interleaving manner. By adjusting the peak limit of the inductor current for a particular output (assuming the valley limit is unchanged), the energy being transferred from L to that output can be independently controlled. The underlying control scheme of the SITO inverter will be discussed shortly.

For a SIMO inverter with n outputs, the switching frequency of S_{main} (f_{sw}) is n times greater than the resonant frequency of the LC resonant tank (f_o). Mathematically, f_{sw} can be expressed as

$$f_{sw} = nf_o = \frac{n\omega_o}{2\pi} = \frac{n}{2\pi\sqrt{L_oC_o}} \qquad (5.8)$$

where ω_o is the resonant frequency in radians, and L_o and C_o are the values of the resonant inductor and resonant capacitor, respectively.

A real application of SIMO inverter is to drive multiple transmitting coils in a multi-device wireless charger. Hence, the chosen resonant frequency should fall within a valid frequency band specified by the WPT standard. For example, the Qi standard defines a range of operating frequencies between 110 and 205 kHz for low-power Qi chargers up to 5 W and 80–300 kHz for medium-power Qi chargers up to 120 W [5,6]. After the resonant frequency is chosen, the proper values of L_o and C_o in the output resonant tank can therefore be determined.

5.3.3 Control Scheme

Besides the circuit topology, another major consideration of the SIMO inverter is to come up a simple, stable, and scalable control scheme for regulating the inductor current flowing into each individual output. This makes conventional hysteresis control a seemingly attractive option due to its unconditional stability. Unfortunately, a big disadvantage of current control using traditional fixed-band hysteresis control is variable switching frequency. This results in a large amount of harmonic contents (or distortions) in the output voltage or current, which will complicate the design of the output filter. Also, the switching frequency depends on various factors such as the operating point, the feedback ripple, the component parasitic elements, and the switching delays. It would be rather difficult to predict the frequency due to the complex interactions of these factors. Hence, hysteretic current control is deemed not suitable for practical Qi-compliant products that need to pass the electromagnetic compatibility (EMC) and electromagnetic interference compliance tests. Consequently, a fixed-frequency hysteretic current control, also referred to as quasi-hysteretic control, is proposed [16]. Figure 5.5 shows the circuit diagram of the power stage of the buck-boost SITO inverter with the quasi-hysteretic control logic. Figure 5.6 shows the ideal timing diagram of the key signals of this inverter.

The quasi-hysteretic control scheme requires a master clock signal (mclk) to achieve fixed switching frequency. The frequency of this clock signal is equal to the switching frequency (f_{sw}) of the SITO inverter. By default, the PCCM enable signal (pccm_en) is always at logic 1 (high) when the SITO inverter operates in PCCM. In DCM, it is set to logic 0 (low). A standard three-phase nonoverlapping clock generator can be used to generate the three output enables signals (outen1, outen2, outen3). To start with, the first output is enabled (outen1 = 1), while the second and third outputs are disabled (outen2 = 0 and outen3 = 0). At the rising edge of the clock signal, the main switch (S_{main}) is turned ON to start the charging phase. A fast comparator is used to compare the sensed inductor current ($I_{L, sen}$) against the peak current limit of the first output ($I_{L, peak1}$). As the main inductor is charged up, $I_{L, sen}$ continues to rise. This charging phase is represented by state (1–1) in Figure 5.6. The peak crossing of the inductor current will finally occur when $I_{L, sen}$ is equal to $I_{L, peak1}$. At this time, S_{main} will be turned OFF, and the first output switches (S_{out1a}, S_{out1b}) are turned ON. This peak-crossing event marks the transition of the SITO inverter from Mode 1 to Mode 2. As the main inductor releases its storage

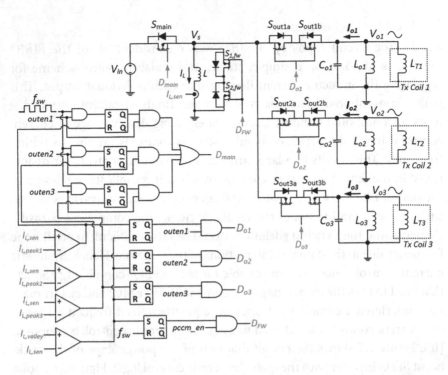

FIGURE 5.5 Circuit diagram of the buck-boost SITO inverter with quasi-hysteretic control logic.

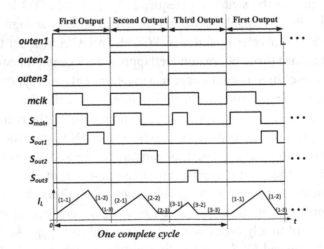

FIGURE 5.6 Ideal timing diagram of the key signals of the buck-boost SITO inverter.

energy to the first output, the inductor current continues to decrease. This discharging phase is represented by state (1–2) in Figure 5.6. The valley crossing of the inductor current will eventually occur when $I_{L,\,sen}$ is equal to the valley current limit ($I_{L,\,valley}$). At this time, S_{out1a} and S_{out1b} are turned OFF, while the freewheeling switches ($S_{1,fw}$, $S_{2,fw}$) are turned ON. This valley-crossing event triggers the SITO inverter to make a transition from Mode 2 to Mode 3. The SITO inverter will remain idle in Mode 3 until the first output is disabled ($outen1 = 0$). Now, the second output is enabled ($outen2 = 1$), while the third output remains disabled ($outen3 = 0$). At the next rising edge of the master clock, the main switch (S_{main}) is turned ON to restart the charging phase. The same control sequence is then repeated for the second output, followed by the third output.

5.4 MATHEMATICAL PROOF OF SINUSOIDAL OSCILLATION

An important property of the buck-boost SIMO inverter is the generation of sinusoidal output voltages. In this section, this property is formally investigated. A rigorous mathematical proof of sinusoidal oscillation in each mode of operation will be provided. It aims to show that this new type of SIMO inverter can theoretically produce sinusoidal oscillations at the outputs. Without loss of generality, the first output of the buck-boost SITO inverter is considered in the theoretical analysis. Since the equivalent circuit for each output is virtually the same in each mode of operation, the same method of analysis can be repeated for the second (or third) output.

5.4.1 Mode 1-Proof of Sinusoidal Oscillation

Figure 5.7 shows the equivalent circuit of the SITO inverter operating in Mode 1 for the first output. Since the output switch (S_{out1}) is opened, the output resonant tank is fully decoupled from the DC input voltage (V_{in})

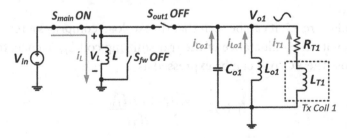

FIGURE 5.7 Equivalent circuit of the SITO inverter operating in Mode 1 for the first output.

and the main inductor (L). Notice that the branch current in the resonant tank and transmitting coil (i.e., i_{Co1}, i_{Lo1}, and i_{T1}) flows from ground to the output node (V_{o1}). By invoking Kirchhoff's current law (KCL) at the output node, we have

$$i_{Co1}(t) + i_{Lo1}(t) + i_{T1}(t) = 0 \tag{5.9}$$

Equation (5.9) can be transformed into the frequency domain (or s-domain) as follows.

$$I_{Co1}(s) + I_{Lo1}(s) + I_{T1}(s) = 0 \tag{5.10}$$

Since C_{o1} and L_{o1} are connected in parallel, $V_{co1}(t) = V_{Lo1}(t) = -V_{o1}(t)$. The current across the resonant capacitor $i_{co}(t)$ can be expressed as

$$I_{Co1}(t) = C_{o1} \frac{dv_{Co1}(t)}{dt} = -C_{o1} \frac{dv_{o1}(t)}{dt} \tag{5.11}$$

By applying Laplace transform to equation (5.11), we have

$$I_{Co1}(s) = C_{o1}\left[sV_{Co1}(s) - V_{Co1} \right] = -C_{o1}\left[sV_{01}(s) + V_{o1} \right] \tag{5.12}$$

where V_{co1} and V_{o1} are the initial values of the resonant capacitor voltage and the output voltage (i.e., at time $= t_0$), respectively. In other words, $V_{co1} = v_{co1}(t_0)$ and $V_{o1} = v_{o1}(t_0)$.

Likewise, the current through the resonant inductor in the s-domain can be written as

$$I_{Lo1}(s) = \frac{V_{Lo1}(s)}{sL_{o1}} + \frac{I_{Lo1}}{s} = -\frac{V_{o1}(s)}{sL_{o1}} + \frac{I_{Lo1}}{s} \tag{5.13}$$

where I_{Lo1} is the initial value of the resonant inductor current, i.e., $I_{Lo1} = i_{Lo1}(t_0)$.

From Figure 5.7, it can be seen that the AC load comprises a resistor R_{T1} in series with the inductance of the transmitting coil L_{T1}. Hence, the current across the AC load can be expressed as

$$I_{T1}(s) = \frac{-V_{o1}(s) + L_{T1}I_{To1}}{R_{T1} + sL_{T1}} \tag{5.14}$$

where I_{To1} is the initial value of the current across L_{T1} at time $= t_0$, i.e., $I_{To1} = i_{T1}(t_0)$.

Now, by substituting Equations (5.12)–(5.14) into (5.10), we have

$$-C_{o1}\left[sV_{o1}(s)+V_{o1}\right]=\frac{V_{o1}(s)}{sL_{o1}}+\frac{I_{Lo1}}{s}+\frac{L_{T1}I_{To1}-V_{o1}(s)}{R_{T1}+sL_{T1}}=0 \qquad (5.15)$$

By rearranging the terms in Equation (5.15) to solve for $V_{o1}(s)$, we have

$$V_{o1}(s)=\frac{-sC_{o1}V_{o1}L_{o1}\left(R_{T1}+sL_{T1}\right)+L_{o1}I_{Lo1}\left(R_{T1}+sL_{T1}\right)+sL_{o1}L_{T1}I_{To1}}{s^2L_{o1}C_{o1}\left(R_{T1}+sL_{T1}\right)+s\left(L_{o1}+L_{T1}\right)+R_{T1}} \qquad (5.16)$$

Suppose $L_{T1} \gg L_{o1}$. Equation (5.16) can be reduced to

$$V_{o1}(s)=K_1\frac{s}{s^2+\omega_o{}^2}+K_2\frac{1}{s^2+\omega_o{}^2}+K_3\frac{1}{s+\dfrac{R_T}{L_T}} \qquad (5.17)$$

where

$$K_1=-V_{o1}+\frac{R_{T1}L_{o1}L_{T1}I_{To1}}{R_{T1}{}^2L_{o1}C_{o1}L_{T1}{}^2} \qquad (5.18a)$$

$$K_2=\frac{I_{Lo1}}{C_{o1}}+\frac{L_{T1}{}^2I_{To1}}{C_{o1}\left(R_{T1}{}^2L_{o1}C_{o1}L_{T1}{}^2\right)} \qquad (5.18b)$$

$$K_3=\frac{-R_{T1}L_{o1}L_{T1}I_{To1}}{R_{T1}{}^2L_{o1}C_{o1}L_{T1}{}^2} \qquad (5.18c)$$

and

$$\omega_o=\frac{1}{\sqrt{L_oC_o}} \qquad (5.18d)$$

By applying inverse Laplace transform, the output voltage in time domain will appear as follows.

$$v_{o1}(t)=a_1.\cos(\omega_o t)+b_1\sin(\omega_o t)+c_1e^{-\left(\frac{R_{T1}}{L_{T1}}\right)t}$$

$$\qquad (5.19)$$

$$=a_1\cos(\theta_o)+b_1\sin(\theta_o)+c_1e^{-\left(\frac{R_{T1}}{L_{T1}}\right)t}$$

where $a_1 = K_1$, $b_1 = K_2\omega_o^{-1}$, $c_1 = K_3$, and $\theta_1 = \omega_o t$.

Note that under steady-state condition, the absolute value of t becomes very large. Thus, the value of $e^{-\frac{R_{T1}}{L_{T1}}t}$ tends toward zero, which means that the last term in equation (5.19) drops out. Hence, equation (5.19) can be further reduced to the following form.

$$v_{o1}(t) = a_1 \cos(\theta_1) + b_1 \sin(\theta_1) \tag{5.20}$$

Let $\sin(\beta_1) = \dfrac{a_1}{\sqrt{a_1^2 + b_1^2}}$ and $\cos(\beta_1) = \dfrac{b_1}{\sqrt{a_1^2 + b_1^2}}$. Equation (5.20) can be reexpressed as

$$v_{o1}(t) = \left(\sqrt{a_1^2 + b_1^2}\right)\left[\cos(\theta_1)\sin(\beta_1) + \sin(\theta_1)\cos(\beta_1)\right] = \left(\sqrt{a_1^2 + b_1^2}\right)\sin(\theta_1 + \beta_1)$$

$$= \left(\sqrt{a_1^2 + b_1^2}\right)\sin(\omega_o t + \beta_1) \tag{5.21}$$

Equation (5.21) shows that the first output voltage $v_{o1}(t)$ is a pure sinusoidal signal with the same frequency as the resonant frequency of the LC resonant tank.

5.4.2 Mode 2-Proof of Sinusoidal Oscillation

Figure 5.8 shows the equivalent circuit of the inverter in Mode 2 for the first output. Unlike its boost-only predecessor, the DC input voltage

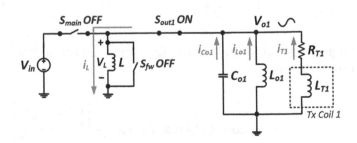

FIGURE 5.8 Equivalent circuit of the SITO inverter operating in Mode 2 for the first output.

(V_{in}) is disconnected from the output node (V_{o1}) in Mode 2 because of the opening of S_{main}. By invoking Kirchhoff's voltage law around the mesh (loop) formed by L and the LC resonant tank, we can write

$$v_L(t) + v_{o1}(t) = 0 \qquad (5.22)$$

where $v_L(t)$ is the instantaneous voltage across the main inductor L, and $v_{o1}(t)$ is the instantaneous output voltage in the time domain.

By applying Laplace transform, equation (5.22) can be rewritten as

$$L\left[sI_L(s) - I_{L,peak}\right] + V_{o1}(s) = 0 \qquad (5.23)$$

where $I_L(s)$ is the inductor current, $I_{L, peak}$ is the peak value of the inductor current, and $V_{o1}(s)$ is the output voltage in the frequency domain.

Since $I_{L,peak} = \dfrac{D_{11}T_sV_{in}}{L}$, equation (5.23) can be rewritten as

$$sLI_L(s) - D_{11}T_sV_{in} + V_{o1}(s) = 0 \qquad (5.24)$$

where D_{11} is the on-time duty ratio corresponding to the first output in Mode 1.

Now, by invoking KCL at the common ground (as depicted in Figure 5.8), we have

$$i_L(t) = i_{Co1}(t) + i_{Lo1}(t) + i_{T1}(t) \qquad (5.25)$$

where $i_L(t)$ is the inductor current, $i_{Co1}(t)$ is the current across the resonant capacitor C_{o1}, $i_{Lo1}(t)$ is the current across the resonant inductor L_{o1}, and $i_{T1}(t)$ is the current across the transmitter coil L_{T1}.

By applying Laplace transform to equation (5.25), we have

$$I_L(s) = C_{o1}\left[sV_{o1}(s) - V_{o1}\right] + \frac{V_{o1}(s)}{sL_{o1}} + \frac{I_{Lo1}}{s} + \frac{V_{o1}(s) + L_{T1}I_{To1}}{L_{T1}s + R_{T1}} \qquad (5.26)$$

where $V_{o1} = v_{o1}(t_1)$, $I_{Lo1} = I_{Lo1}(t_1) = I_{L, peak}$, $I_{To1} = i_{T1}(t_1)$, and $t_1 = t_0 + D_{11}T_s$.

By substituting equation (5.26) into (5.24) and rearranging, $V_{o1}(s)$ can be solved as follows.

$$V_{o1}(s) = \frac{1}{LC_{o1}} \left(V_{o1} - \frac{L_{T1}I_{To1}}{C_{o1}(R_{T1}+sL_{T1})} \right) \left(\frac{s}{s^2 + \frac{1}{LC_{o1}} + \frac{1}{L_{o1}C_{o1}}} \right)$$

$$+ \frac{1}{LC_{o1}} \left(\frac{I_{L,peak}}{C_{o1}} - \frac{I_{Lo1}}{C_{o1}} \right) \left(\frac{1}{s^2 + \frac{1}{LC_{o1}} + \frac{1}{L_{o1}C_{o1}}} \right) \tag{5.27}$$

Suppose $\omega_1^2 = \frac{1}{LC_{o1}} + \frac{1}{L_{o1}C_{o1}}$. Equation (5.27) can be expressed in terms of ω_1 as follows.

$$V_{o1}(s) = \frac{V_{o1}}{LC_{o1}} \left(\frac{s}{s^2 + \omega_1^2} \right) + \frac{1}{LC_{o1}^2} \left(I_{L,peak} - I_{Lo1} \right) \left(\frac{1}{s^2 + \omega_1^2} \right)$$

$$- \frac{L_{T1}I_{T01}}{LC_{o1}^2} \left(\frac{1}{R_{T1}+sL_{T1}} \right) \left(\frac{s}{s^2 + \omega_1^2} \right) \tag{5.28}$$

By performing partial fraction expansion on the last term on the R.H.S. of equation (5.28) and rearranging, we have

$$V_{o1}(s) = \frac{V_{o1}}{LC_{o1}} \left(\frac{s}{s^2 + \omega_1^2} \right) + \frac{1}{LC_{o1}^2} \left(I_{L,peak} - I_{Lo1} - I_{To1} \right) \left(\frac{1}{s^2 + \omega_1^2} \right)$$

$$+ \frac{I_{To1}R_{T1}L_{o1}}{C_{o1}(L+L_{o1})} \left(\frac{1}{R_{T1}+sL_{T1}} \right) \tag{5.29}$$

Now, by applying inverse Laplace transform to both sides of (5.29), we have

$$v_{o1}(t) = a_2 \cos(\theta_2) + b_2 \sin(\theta_2) + c_2 e^{-\left(\frac{R_{T1}}{L_{T1}} \right)t} \tag{5.30}$$

where

$$a_2 = \frac{V_{o1}}{LC_{o1}}, \tag{5.31a}$$

$$b_2 = \frac{1}{\omega_1 L C_{o1}{}^2} \left(I_{L,peak} - I_{Lo1} - I_{To1} \right), \tag{5.31b}$$

$$c_2 = \frac{I_{To1} R_{T1} L_{o1}}{L_{T1} C_{o1} (L + L_{o1})}, \tag{5.31c}$$

and

$$\theta_2 = \omega_1 t. \tag{5.31d}$$

In steady-state condition, the absolute value of t becomes very large, and hence, the exponential term $e^{\frac{R_T}{L_T} t}$ tends to zero, which implies that the third term in (5.30) can be neglected.

Thus, equation (5.30) becomes

$$v_{o1}(t) = a_2 \cos(\theta_2) + b_2 \sin(\theta_2) \tag{5.32}$$

Let $\sin(\beta_2) = \dfrac{a_2}{\sqrt{a_2{}^2 + b_2{}^2}}$ and $\cos(\beta_2) = \dfrac{b_2}{\sqrt{a_2{}^2 + b_2{}^2}}$. Equation (5.32) can be rewritten in the following form.

$$v_{o1}(t) = \left(\sqrt{a_2{}^2 + b_2{}^2} \right) \left[\cos(\theta_2) \sin(\beta_2) + \sin(\theta_2) \cos(\beta_2) \right] = \left(\sqrt{a_2{}^2 + b_2{}^2} \right) \sin(\theta_2 + \beta_2)$$

$$= \left(\sqrt{a_2{}^2 + b_2{}^2} \right) \sin(\omega_1 t + \beta_2) \tag{5.33}$$

Hence, it is proven that in Mode 2, the output voltage $v_{o1}(t)$ is a pure sinusoidal signal whose frequency is given by $\omega_1 = \sqrt{\dfrac{1}{L C_{o1}} + \dfrac{1}{L_{o1} C_{o1}}}$. In particular, when $L \gg L_{o1}$, $\omega_1 \approx \omega_0$. In other words, the frequency of the sinusoidal output voltage in Mode 2 is approximately equal to the resonant frequency of the LC resonant tank when the value of the main inductor L is much greater than that of the resonant inductor L_{o1}.

5.4.3 Mode 3-Proof of Sinusoidal Oscillation

Figure 5.9 shows the equivalent circuit of the inverter in Mode 3 for the first output. Both S_{main} and S_{out1} are switched OFF, which means that the resonant tank is completely isolated from the main inductor L.

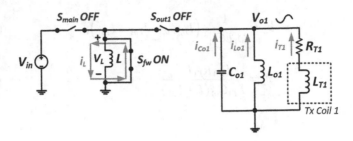

FIGURE 5.9   Equivalent circuit of the SITO inverter operating in Mode 3 for the first output.

Despite the presence of a free-wheeling current across the inductor (in PCCM), the circuit analysis in Mode 3 is largely identical to that in Mode 1. A subtle difference between the two modes is that the initial values for the output voltage and the resonant current in Mode 3 are obtained at time t_2 (instead of t_0), as depicted in Figure 5.4a. Nonetheless, the proof of sinusoidal oscillation in Mode 1 remains valid also for Mode 3. It follows that the frequency of the sinusoidal output voltage in Mode 3 is identical to that in Mode 1, i.e., $\omega_o = \dfrac{1}{\sqrt{L_{o1}C_{o1}}}$.

5.5   EXTENSION FROM SITO TO SIMO

A key property of the SIMO-based inverter architecture is its extension (or scalability) to N number of AC outputs. In practice, only a finite number of AC outputs can be implemented in a SIMO inverter. This is mainly attributed to the maximum possible energy stored in the main inductor as well as the chosen resonant frequency. A theoretical maximum number of AC outputs that can be realized by the SIMO inverter architecture will be formally investigated.

In principle, to maximize the number of outputs at a given resonant period (T_o), the SIMO inverter needs to operate at the boundary condition between PCCM and CCM with no freewheeling period. Figure 5.10 illustrates the timing diagram of the inductor current of the SIMO inverter in this boundary condition.

By simple geometry, due to the disappearance of the idle period, the inductor current (I_L), as shown in Figure 5.10, attains its maximum peak value at a particular switching frequency (T_s), where $T_s = T_o/N$. For a given DC offset of the inductor current ($I_{L,DC}$), maximum storage energy in the main inductor can therefore be transferred sequentially to each output. For ease of

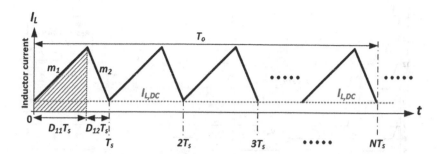

FIGURE 5.10   Timing diagram of the inductor current of the SIMO inverter in BCM.

illustration, the SIMO inverter is assumed to operate in balanced load condition under which the same amount of energy is delivered from the main inductor to each output. In Mode 1, the input current can be expressed as

$$I_{in} = \overline{I_L} = \frac{1}{T_o} \int_{nT_s}^{(n+D_{on})T_s} i_L(t)dt \qquad (5.34)$$

where n is a positive integer ranging from 0 to $(N-1)$, and D_{on} is the on-time duty ratio of S_{main}.

Since the inductor current has uniform peak values for all outputs in balanced load condition, the value of D_{on} is identical across all outputs (i.e., $D_{on} = D_{11} = D_{21} = \ldots = D_{N1}$). Note that the integral in equation (5.34) represents the blue shaded area shown in Figure 5.10. Hence, equation (5.34) can be reexpressed as

$$I_{in} = \overline{I_L} = \frac{V_{in}D_{on}^2 T_s^2 + 2D_{on}LI_{L,dc}T_s}{2LT_o} \qquad (5.35)$$

At first, an ideal (lossless) SIMO inverter is examined. The input power P_{in} can be written as

$$P_{in} = V_{in}I_{in} = V_{in}\overline{I_L} = \frac{V_{in}^2 D_{on}^2 T_s^2 + 2V_{in}D_{on}LI_{L,dc}T_s}{2LT_o} \qquad (5.36)$$

Since $P_{in}=P_{out}$ for a lossless inverter (where P_{out} is the output power per channel), we can write

$$P_{in} = P_{out} = \frac{V_{in}^2 D_{on}^2 T_s^2 + 2V_{in}D_{on}LI_{L,dc}T_s}{2LT_o} \qquad (5.37)$$

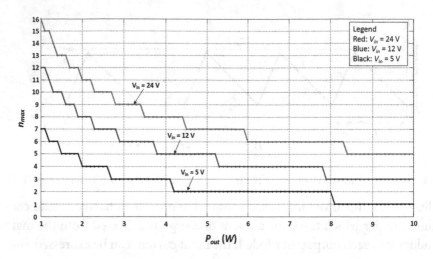

FIGURE 5.11   Plot of the theoretical maximum number of outputs (n_{max}) versus output power per channel (P_{out}) of the buck-boost SIMO inverter.

By re-arranging the terms in equation (5.37) and replacing T_s with T_o/n_{max}, the following quadratic equation can be obtained.

$$2LP_{out}n_{max}^2 - 2V_{in}D_{on}LI_{L,DC}n_{max} - V_{in}^2D_{on}^2T_o = 0 \tag{5.38}$$

where n_{max} is the maximum possible number of outputs.

The discriminant Δ of equation (5.38) is given by

$$\Delta = 4D_{on}^2V_{in}^2\left(L^2I_{L,dc}^2 + 2LP_{out}T_o\right) > 0 \tag{5.39}$$

Since Δ must be a positive value, it implies that the quadratic equation in (5.38) has two real roots (r_1, r_2), which can be represented as

$$r_1,r_2 = \frac{D_{on}V_{in}I_{L,dc}\left(1\pm\sqrt{1+\dfrac{2P_{out}T_o}{LI_{L,dc}^2}}\right)}{2P_{out}} \tag{5.40}$$

Because n_{max} must be a positive integer, the negative root is therefore eliminated. Hence, the theoretical maximum number of SIMO inverter can be obtained as follows.

$$n_{max} = floor\left(\frac{D_{on}V_{in}I_{L,dc}\left(1+\sqrt{1+\dfrac{2P_{out}T_o}{LI_{L,dc}^2}}\right)}{2P_{out}}\right) \tag{5.41}$$

It should be noted that this equation is applicable to *all* types of SIMO inverter, including the buck-boost SIMO inverter. In periodic steady state of such inverter, the net change of the inductor current (I_L) is zero. Thus, by invoking volt-second balance, we have

$$m_1 D_{on}T_s = m_2(1-D_{on})T_s \tag{5.42}$$

Since $m_1 = V_{in}/L$ and $m_2 \cong V_m/L$, the on-time duty ratio D_{on} can be expressed as

$$D_{on} = \frac{V_m}{V_{in}+V_m} \tag{5.43}$$

where V_m is the amplitude of the sinusoidal output voltage and V_{in} is the DC input voltage.

By substituting equation (5.43) into (5.41), n_{max} can be expressed in terms of V_m and V_{in} as follows.

$$n_{max} = floor\left(\frac{V_{in}V_m I_{L,dc}\left(1+\sqrt{1+\dfrac{2P_{out}T_o}{LI_{L,dc}^2}}\right)}{2(V_{in}+V_m)P_{out}}\right) \tag{5.44}$$

Equation (5.44) defines the theoretical maximum number of outputs of the SIMO inverter. Let us assume that each output of the SIMO inverter is connected to a pure resistive load (R_L); the real power P_{out} can be expressed as a function of V_m and R_L as follows.

$$P_{out} = \frac{V_{o,rms}^2}{R_L} = \frac{V_m^2}{2R_L} \tag{5.45}$$

where $V_{o,rms}$ is the RMS value of the output voltage and $V_{o,rms} = \dfrac{V_m}{\sqrt{2}}$.

From equation (5.45), given a particular output power P_{out}, the value of V_m can be obtained easily. Hence, the values of all parameters in equation (5.44) are known beforehand, and so, n_{max} of a particular SIMO inverter can be determined.

As an example, the scalability of the buck-boost SIMO inverter is investigated using these parameter values: $L=6$ μH, $R_L=50$ Ω, $I_{L,\,dc}=2$ A, and $T_o=10$ μs. Based on equation (5.44), we can determine the relationship between n_{max} and P_{out} for $V_{in}=5$, 12 and 24 V, which is graphically represented in Figure 5.11. Intuitively, because the maximum amount of energy being stored in the main inductor is constant, an increase in the output power across channels means that a fewer number of output channels can be achieved by the SIMO inverter. In addition, an increase in the input voltage V_{in} (or input power P_{in}) at a given output power will lead to an increase in the total achievable number of outputs of the SIMO inverter. Figure 5.11 shows an inverse relationship between the theoretical maximum number of outputs and the output power per channel. Indeed, this figure is beneficial to a practical design of a SIMO inverter as the maximum achievable number of outputs can be predicted well in advance.

From a practical point of view, the presence of parasitic resistances, e.g., the ESR of the inductor/capacitor, the on-resistance of the power switch, in a real circuit of SIMO inverter degrades the power conversion efficiency (η), where $P_{out}=\eta P_{in}$ and $0<\eta<1$. By applying $P_{in} = \dfrac{P_{out}}{\eta}$ to equation (5.36) and then solving for n_{max}, we have

$$n_{max} = floor\left(\frac{\eta V_{in} V_m I_{L,dc}\left(1+\sqrt{1+\dfrac{2P_{out}T_o}{\eta L I_{L,dc}^2}}\right)}{2(V_{in}+V_m)P_{out}}\right) \qquad (5.46)$$

Equation (5.46) represents the scalability model of a practical SIMO inverter, which accounts for the power efficiency. At a given output power, n_{max} decreases with a smaller power efficiency. Figure 5.12 shows a three-dimensional plot of n_{max} versus the output power per channel (P_{out}) and power efficiency (η). As can be seen in Figure 5.12, at a given output power per channel, an increasing number of outputs can be achieved in the SIMO inverter by increasing the power efficiency. For instance, at an output power of 6 W per channel, the maximum achievable number of outputs is increased from 3 to 4 when the power efficiency is increased

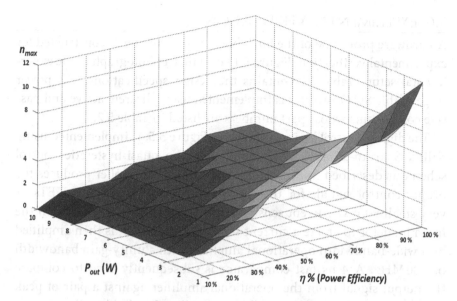

FIGURE 5.12 Three-dimensional plot showing the relationship between the theoretical maximum number of outputs (n_{max}) versus the output power per channel (P_{out}) and power efficiency (η).

from 80% to 90% or above. Likewise, at a particular power efficiency, a larger number of outputs can be realized by lowering the output power per channel. As an example, the maximum number of outputs of a SIMO inverter with a power efficiency of 85% is increased from 3 to 5 when the output power per channel is reduced from 9 to 4 W.

In particular, with zero DC offset in the inductor current (i.e., $I_{L, dc}=0$), the SIMO inverter operates in boundary conduction mode (BCM), which is also referred to as critical conduction mode. In BCM, the inverter operates at the boundary between continuous conduction mode (CCM) and discontinuous conduction mode (DCM) without the freewheeling (idle) period. By substituting $I_{L, dc}=0$ and $P_{in} = \dfrac{P_{out}}{\eta}$ into equation (5.36), n_{max} can be obtained as

$$n_{max} = floor\left(\frac{V_{in}V_m}{(V_{in}+V_m)} \sqrt{\frac{\eta T_o}{2LP_{out}}} \right) \qquad (5.47)$$

Essentially, equation (5.47) defines the maximum number of AC outputs of the SIMO inverter operating in BCM.

5.6 EXPERIMENTAL VERIFICATION

A hardware prototype of the buck-boost SITO inverter is constructed for experimental verification. Figure 5.13 contains a photograph of the experimental setup. Table 5.1 contains the design specifications. The power stage of this SITO inverter is implemented using discrete active and passive components whose part numbers are listed in Table 5.2.

The digital control logic, as shown in Figure 5.5, is implemented using Xilinx Spartan-3E FPGA. By employing the quasi-hysteretic control scheme as described in Section 5.3.3, the digital controller produces the proper control signals for the gate drivers for all power MOSFETs. A very small 50 mΩ current sense resistor is connected in series with the main inductor to create a current sense voltage, which is then amplified by a wide-bandwidth operational amplifier with a unity-gain bandwidth of 200 MHz. A 4-ns fast comparator is subsequently used to compare the output signal from the operational amplifier against a pair of peak threshold (peak current limit of I_L) and valley threshold (valley current limit of I_L) for each output in order to generate the corresponding logic signals for the FPGA.

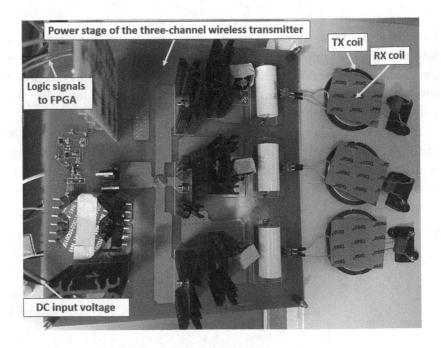

FIGURE 5.13 Photograph of the experimental setup.

TABLE 5.1   Design Specifications of Buck-Boost SITO Inverter

Design Parameter	Value
Input voltage (V_{in})	12 V
Switching frequency (f_{sw})	300 kHz
Output resonant frequency (f_o)	100 kHz
Main inductor (L)	6 μF
ESR of the main inductor (measured at 300 kHz)	90 mΩ
Capacitor in the resonant tank (C_{o1}, C_{o2}, C_{o3})	0.22 μF
ESR of the resonant capacitor (measured at 100 kHz)	30 mΩ
Inductor in the resonant tank (L_{o1}, L_{o2}, L_{o3})	11.5 μH
ESR of the resonant inductor (measured at 100 kHz)	12 mΩ
Load resistor (R_{T1}, R_{T2}, R_{T3})	22 Ω

TABLE 5.2   List of Components with Part Numbers

Component	Part Number
Power stage	
Main inductor (L)	Custom-made
Power MOSFET	IPP083N10N5AKSA1
Gate driver for the power MOSFET	SI8261BAC-C-IS
Current-sensing resistor	MP930
Resonant capacitor	940C10P22K-F
Resonant inductor (L_{o1}, L_{o2}, L_{o3})	Custom-made
Transmitting coil	760308100110
Receiving coil	760308103211
Output resistor (R_{L1}, R_{L2}, R_{L3})	RCH25S22R00JS06
Feedback controller	
Operational amplifier	OPA354
Comparator	AD8611
Xilinx FPGA	Spartan-3E (XC3S250E)

In the first experiment, a balanced load scenario will be verified. Each of the three outputs of the buck-boost SITO inverter is initially connected to a 22 mΩ resistor (i.e., the same resistor value is used across all outputs). Figure 5.14a and b show the measured waveforms of the inductor current (I_L) and the three sinusoidal-like output voltages (V_{o1}, V_{o2}, V_{o3}).

Even though the same set of measured waveforms is shown in both figures, Figure 5.14a highlights the measured RMS values, whereas Figure 5.14b highlights the measured frequencies of the three outputs. As can be seen clearly in both figures, the SITO inverter is capable of producing three sinusoidal-like output voltages simultaneously. The RMS values

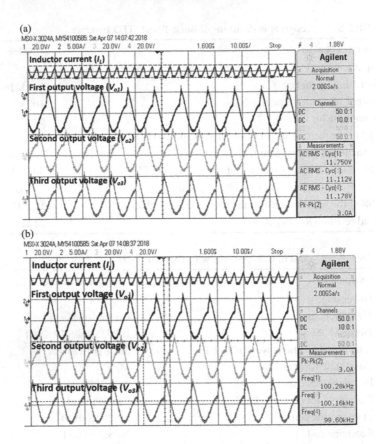

FIGURE 5.14 Measured waveforms of the inductor current I_L and the three output voltages (V_{o1}, V_{o2}, V_{o3}) of the buck-boost SITO inverter with (a) comparable RMS values and (b) the same frequency in a balanced load scenario.

of the three output voltages are measured to be 11.750, 11.112, and 11.178 V, respectively. Also, the measured output voltages have the same frequency of around 100 kHz and a phase difference of 120° between two adjacent outputs. In summary, it is experimentally verified that the SITO inverter can operate properly in balanced load condition by generating three nearly identical sinusoidal-like output voltages of the same amplitude and frequency. No voltage spikes are observed in the measured waveforms. Some distortions are seen in the output voltage waveforms only when the inverter traverses from Mode 1 to Mode 2. This is largely attributed to a jump in the frequency of the output voltage during this mode transition, as explained in Section 5.4. On the other hand, the measured inductor current has a positive DC offset and a peak-to-peak current ripple of 3.0 A. This verifies that the SITO inverter operates properly in PCCM.

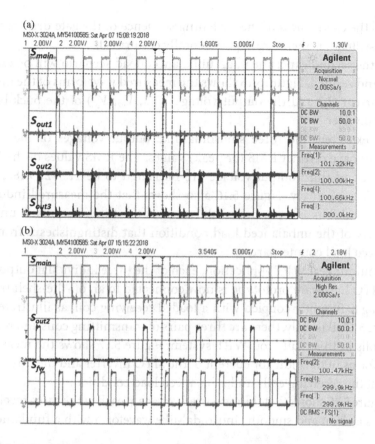

FIGURE 5.15 Measured waveforms of (a) the switching sequence for the gate drive signals of the main switch (S_{main}) and the three output switches (S_{out1}, S_{out2}, S_{out3}) and (b) the switching sequence for the gate drive signals of the main switch (S_{main}), the second output switch (S_{out2}), and the freewheeling switch (S_{fw}).

Figure 5.15a shows the actual switching sequence of the gate drive signals of the main switch (S_{main}) and the output switches (S_{out1}, S_{out2}, S_{out3}). It is experimentally verified that S_{main} switches at a frequency of 300 kHz, which is *three times* faster than that of the output switch. Because a four-channel oscilloscope is used, at most, four separate waveforms can be captured at a time. So, the gate drive signal of the freewheeling switch (S_{fw}) has to be shown in another figure. Figure 5.15b shows the switching sequence of S_{main}, S_{out2}, and S_{fw}. It shows that both S_{main} and S_{fw} switch at the same frequency because they need to be turned ON in every switching cycle. Shortly after S_{main} is switched OFF, one of the output switches ($S_{out1}/S_{out2}/S_{out3}$) is switched ON. Likewise, S_{fw} is switched ON immediately after the output switch is switched OFF, and this switching pattern repeats in every cycle. The correctness of the switching sequence of the gate drive signals of *all* switches is therefore experimentally confirmed.

In the second experiment, an unbalanced load scenario will be examined and verified. Figure 5.16 shows the measured waveforms of the inductor current (I_L) and the output voltages (V_{o1}, V_{o2}, V_{o3}) of the buck-boost SITO inverter.

As expected, the three sinusoidal-like output voltages of the SITO inverter have distinct RMS (or peak) values. The RMS values of the first, second, and third output voltage are measured to be 11.372, 9.185, and 7.975 V, respectively. Figure 5.16 also shows that the measured inductor current exhibits uneven peak values across the three outputs, a unique property of the unbalanced load condition that distinguishes it from its balanced load counterpart.

To mimic real WPT application, each of the three individual outputs of the SITO hardware prototype is now connected to an off-the-shelf transmitting coil tightly coupled with a loaded receiving coil, as illustrated in Figure 5.13. Basically, there are three pairs of transmitting coils and receiving coils in this multi-coil WPT system. Figure 5.17 shows the measured waveforms of the inductor current of the SITO inverter and the three voltages at the receiving coils under balanced load condition.

Figure 5.17 shows that the measured voltage at each of the three receiving coils is a clean and smooth sinusoidal-like waveform with a fundamental

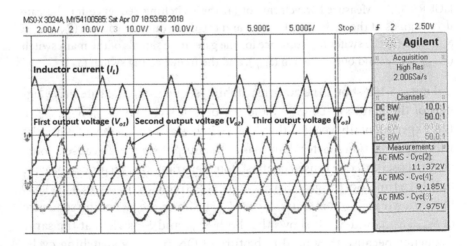

FIGURE 5.16 Measured waveforms of the inductor current (I_L) and the three output voltages (V_{o1}, V_{o2}, V_{o3}) of the buck-boost SITO inverter in an unbalanced load scenario.

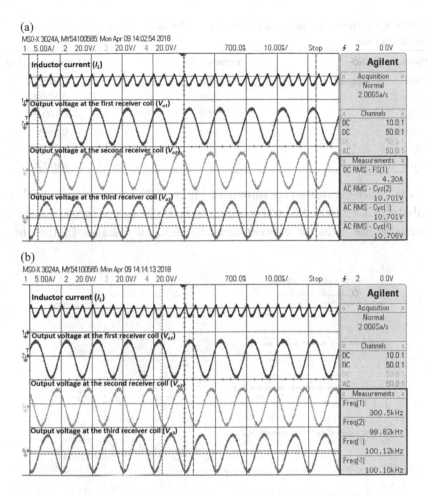

FIGURE 5.17 Measured waveforms of the inductor current of the SITO inverter and the three output voltages at the receiver side showing (a) identical RMS values and (b) same frequencies across the three receiver output channels in balanced load condition.

frequency of around 100 kHz. No noticeable voltage spike is observed in the measured waveform of the voltage induced in each receiving coil. The measured RMS values of the three voltages are 10.701, 10.701, and 10.706 V, which are in very close agreement with each other. This is indicative of a balanced load condition. The harmonic content of the induced voltage in the first receiving coil is also measured. Table 5.3 tabulates the measured RMS values of the fundamental component and the low-order harmonics (up to the sixth order). The total harmonic distortion (THD) is measured to be around 5.87%, which is sufficiently small.

TABLE 5.3 Measured Harmonic Contents of the Induced Voltage at the First Receiving Coil

Harmonic Order	Frequency (kHz)	Measured V_{rms} (V)
First (fundamental)	100	9.975500
Second	200	0.576050
Third	300	0.100464
Fourth	400	0.039917
Fifth	500	0.011597
Sixth	600	0.013794

The experimental setup shown in Figure 5.13 is now configured for unbalanced load condition. Figure 5.18a and b show the measured waveforms of the inductor current and the induced voltages in the three

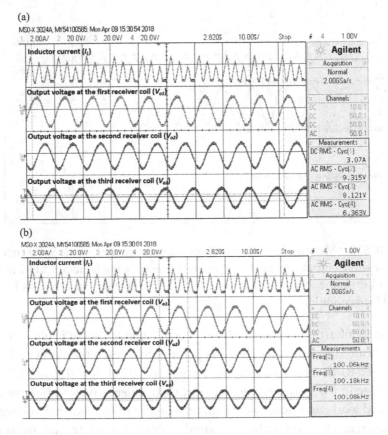

FIGURE 5.18 Measured waveforms of the inductor current and the induced voltages at the three receiving coils showing (a) distinct RMS values and (b) same frequencies under the unbalanced load condition.

receiving coils. Figure 5.18a highlights the fact that the three voltages at the receiver have *unequal* RMS values. Figure 5.18b shows that each of the three voltages continue to oscillate as a clean and smooth sine wave with a fundamental frequency of 100 kHz.

To sum up, the experimental results demonstrate the effectiveness of the buck-boost SITO inverter in producing three sinusoidal-like output voltages with independently controlled peak (or RMS) values from a single DC input voltage. No noticeable cross-regulation is observed when the SITO inverter operates in PCCM. The functionality of the digital controller is also experimentally verified. Basically, the inductor current flowing into each of the three individual output channels can be independently regulated by using the quasi-hysteretic control scheme. Given an input power of 29.88 W, the measured power efficiency of the buck-boost SITO inverter at the rated power of 8.4 W per channel (or a total output power of 25.2 W) is around 84.3%.

5.7 SUMMARY OF KEY POINTS

1. A new family member of the SIMO inverter, known as the buck-boost SIMO inverter is introduced. Unlike its boost-only predecessor, it enables a single-stage DC-AC power conversion from a DC power source into multiple AC output voltages whose RMS values can be greater than or less than the DC input voltage.

2. A quasi-hysteretic control scheme is used in the buck-boost SIMO inverter in order to regulate the main inductor current between the peak current limit and the valley current limit. Unlike conventional hysteretic control with variable frequency, the quasi-hysteretic control scheme allows the SIMO inverter to operate at a fixed switching frequency while still carrying the benefit of unconditional stability.

3. By operating the SIMO inverter in PCCM, a positive DC offset is added to the inductor current. This helps increase the energy stored in the main inductor in the charging phase. Thus, more energy can be subsequently delivered from the main inductor to the output in the discharging phase, thereby increasing the output power.

4. The scalability of the SITO inverter to SIMO is formally investigated. This leads to the discovery of a general mathematical model for predicting the maximum achievable number of AC outputs in the SIMO inverter.

5. A practical application of the buck-boost SIMO inverter is that in a multi-device WPT system, it can act as a multi-channel wireless power transmitter (or wireless charger) with low-to-medium power ratings. It enables fast, convenient, and simultaneous charging of multiple unidentical electronic devices.

REFERENCES

1. M. Pinuela, D. C. Yates, S. Lucyszyn, and P. D. Mitcheson, "Maximizing DC-to-load efficiency for inductive power transfer", *IEEE Trans. Power Electron.*, vol. 28, no. 5, pp. 2437–2447, May 2013.
2. D. Liu, H. Hu, and S. V. Georgakopoulos, "Misalignment sensitivity of strongly coupled wireless power transfer system", *IEEE Trans. Power Electron.*, vol. 32, no. 7, pp. 5509–5519, Jul. 2017.
3. A. K. RamRakhyani and G. Lazzi, "On the design of efficient multi-coil telemetry system for biomedical implants", IEEE Trans. Biomed. Circuits Syst., vol. 7, no. 1, pp. 11–23, Feb. 2013.
4. J. P. W. Chow, H. S. H. Chung, C. S. Cheng; A. Gungor, S. C. Tang, and L. L. H. Chan, "Modeling and experimentation of loosely-coupled coils with transmitter having orthogonally-placed windings", *2015 IEEE Energy Conversion Congress and Exposition (ECCE)*, pp. 4927–4934, Sep. 2015.
5. The Qi Wireless Power Transfer System Power Class 0 Specification, Part 4: Reference Designs, *Wireless Power Consortium*, Version 1.2.4, Jan. 2018.
6. Datasheet: BQ500210, "Qi compliant wireless power transmitter manager," Texas Instruments (2012). [Online] Available: http://www.ti.com/lit/ds/slusal8c/slusal8c.pdf.
7. M. Q. Nguyen, Y. Chou, D. Plesa, S. Rao, and J.-C. Chiao, "Multiple-inputs and multiple- outputs wireless power combining and delivering systems", *IEEE Trans. Power Electron.*, vol. 30, no. 11, pp. 6254–6263, Nov. 2015.
8. B. H. Waters, B. J. Mahoney, V. Ranganathan, and J. R. Smith, "Power delivery and leakage field control using an adaptive phase array wireless power system", *IEEE Trans. Power Electron.*, vol. 30, no. 11, pp. 6298–6309, Nov. 2015.
9. R. Johari, J. V. Krogmeier, and D. J. Love, "Analysis and practical considerations in implementing multiple transmitters for wireless power transfer via coupled magnetic resonance", *IEEE Trans. Ind. Electron.*, vol. 61, no. 4, pp. 1774–1783, Apr. 2014.
10. L. Shi, Z. Kabelac, D. Katabi, and D. Perreault, "Wireless power hotspot that charges all of your devices", *2015 Annual International Conference on Mobile Computing (ACM MobiCom 2015)*, pp. 2–13, Sep. 2015.
11. J. Lee, et al., "Wireless power transmitter and wireless power transfer method thereof in many-to-one communication," U. S. Patent US9306401 B2, 2011.
12. A. H. Mohammadian, et al., "Wireless power transfer using multiple transmit antennas," U. S. Patent US8629650 B2, 2008.

13. User's Guide: WCT1001A/WCT1003A, "WCT1001A/WCT1003A automotive A13 wireless charging application user's guide," NXP Semiconductors (2014). [Online] Available: http://cache.nxp.com/files/microcontrollers/doc/user_guide/WCT100XAWCAUG.pdf.

14. Z. Yao, L. Xiao, and Y. Yan, "A novel multiple output grid-connected inverter based on DSP control", *Proceedings IEEE Power Electronics Specialist Conference*, pp. 317–322, Jun. 2008.

15. A. T. L. Lee, W. Jin, S. Tan, and S. Y. (Ron) Hui, "Buck-boost single-inductor multiple- output high-frequency inverters for medium-power wireless power transfer", *IEEE Trans. Power Electron.*, vol. 34, no. 4, pp. 3457–3473, Apr. 2019.

16. A. Lee, J. Sin, and P. Chan, "Scalability of quasi-hysteretic FSM-based digitally- controlled single-inductor dual-string buck LED driver to multiple string", *IEEE Trans. Power Electron.*, vol. 29, no. 1, pp. 501–513, Jan. 2014.

PROBLEMS

5.1   Draw the circuit diagram of an ideal SIMO noninverting buck-boost inverter.

5.2   List two benefits of using the inverting buck-boost topology rather than its noninverting counterpart to implement the SIMO buck-boost inverter.

5.3   Comment on the advantages and disadvantages of operating the SIMO buck-boost inverter in PCCM.

5.4   A power electronics engineer investigates an ideal SITO inverting buck-boost inverter operating in PCCM with the following design specifications.

Input voltage (V_{in}): 12 V
DC offset of the inductor current ($I_{L, dc}$): 1.5 A
Peak value of the inductor current ($I_{L, pk}$): 2.9 A
Main inductor (L): 10 µH
Resonant inductor (L_o) at each output: 11.5 µH
Resonant capacitor (C_o) at each output: 0.22 µF

a. Calculate the resonant frequency of this SITO buck-boost inverter.

b. Determine the on-time duty ratio of the main switch.

c. Suggest a method to increase the total output power of this SITO inverter without changing the values of the passive components.

5.5 Create a circuit (simulation) model of an ideal SITO inverting buck-boost inverter in Problem 5.4.

a. Suppose the inductor current has a constant DC offset ($I_{L, DC}$) of 1.5 A. Use simulation to determine the maximum achievable peak value of the inductor current ($I_{L, pk, max}$) of the SITO inverter operating in PCCM without cross-regulation.

b. Determine the maximum output power of all outputs by using the result in part (a).

c. Based on the simulated on-time duty ratio of the main switch in part (a), calculate the theoretical maximum output power of all outputs of the SITO buck-boost inverter. Explain the difference between the theoretical and simulated values of the maximum output power.

5.6 A SIFO buck-boost inverter is constructed as shown in Figure 5.19 with the following design parameters. Assume that the SIFO inverter is lossless. Find the maximum total output power of this SIFO inverter.

Input voltage (V_{in}): 24 V
DC offset of the inductor current ($I_{L, dc}$): 2 A
Main inductor (L): 10 µH
Resonant inductor (L_o) at each output: 11.5 µH
Resonant capacitor (C_o) at each output: 0.22 µF
Load resistor at each output: 22 Ω
On-time duty ratio of main switch (D_{main}): 44% (at the maximum output power)

5.7 A practical SIMO inverter incurs conduction losses and switching losses due to the presence of parasitic resistances and capacitances. For a rated output power of 6 W per channel, the power efficiency is 88%. Based on the following specifications, determine the maximum achievable number of outputs of the SIMO inverter.

Input voltage (V_{in}): 12 V
DC offset of the inductor current ($I_{L, dc}$): 1 A
Main inductor (L): 5 µH
Resonant inductor (L_o) at each output: 11.5 µH
Resonant capacitor (C_o) at each output: 0.22 µF
Load resistor at each output: 22 Ω

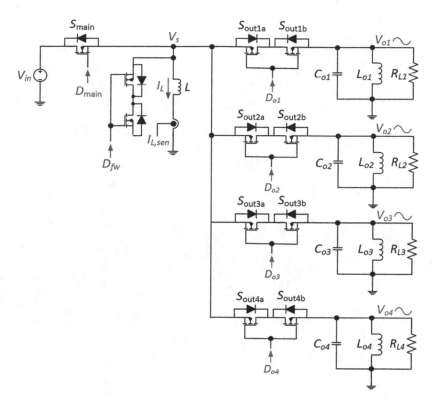

FIGURE 5.19   Single-inductor four-output (SIFO) buck-boost inverter.

CHAPTER 6

Gallium Nitride-Based Single-Inductor Multiple-Output DC-AC Boost Inverters

6.1 INTRODUCTION

Ever since wireless power transfer (WPT) based on coupled magnetic resonance was first proposed by Nicola Tesla over a century ago, significant progress has been made to achieve efficient and cost-effective wireless power transmission. Recently, the WPT technology has quickly emerged as a promising technology in a myriad of practical applications ranging from the charging of low-power electronic devices such as smartphones, tablets, the Internet-of-Things gadgets and medical devices to the charging of high-power electric vehicle. Convenience is perhaps the main reason for most consumers, particularly for travelers, to switch over from traditional wired charging to wireless charging as they no longer need to carry power adapters and cables, given that wireless charging has become a standard amenity at coffee shops, hotels, restaurants, and airports. No charging cables means that mobile users do not need to worry about mismatched connectors or cable compatibility issues due to the lack of a common wired charging standard. Wireless charging also helps cut down

DOI: 10.1201/9781003239833-6

on cable clutter and obviate the need for bulky adapters. Additionally, no repetitive plugging and unplugging of power cables are required in wireless charging, which means less tear and wear on the charging port of a mobile phone. Since the charging port is a common point of failure for mobile devices, wireless charging greatly enhances their reliability and robustness, which may increase the overall lifespan of these devices.

Toward the end of 2014, three competing standards of WPT had emerged, namely Qi (pronounced "chee") standard, Power Matters Alliance (PMA), and Rezence. Qi is an open interface standard that defines wireless charging based on inductive charging over short distances (up to 4 cm or about 1.6 in). The Qi low-power specification was originally published by the Wireless Power Consortium (WPC) in August 2009, which is subsequently renamed as the Qi Baseline Power Profile (BPP) in 2015. Under this specification, a relatively low power (up to 5 W) can be transferred using inductive coupling between two planar coils, which operate at relatively low frequencies ranging from about 110 to 205 kHz. In 2015, WPC formally introduced an extended Qi standard for medium power, which became known as the Qi Extended Power Profile. It supports a maximum power of 15 W with operating frequencies ranging from 80 to 300 kHz. As a matter of fact, a new medium-power wireless power standard is currently in development, which aims at filling the gap between Qi and the Ki Cordless Kitchen standard. Basically, this new medium-power standard is a simple and low-cost solution that delivers wireless power to a wide variety of battery-powered products (e.g. portable power tools, domestic vacuum and lawn mowing robots, drones, electronic bikes, medical devices, etc.) operating in the 30- to 65-W range (and eventually, up to 200 W). Indeed, Qi is now one of the most popular and widely used standards in wireless charging. It has been adopted by various mobile phone OEMs. On the other hand, the mission of PMA, a global, nonprofit, industry organization founded by Procter & Gamble and Powermat Technologies in March 2012, was to formulate a suite of standards for WPT for mobile devices based on inductive coupling technology. This standard is generally targeted at low-power applications with a charging power no more than 5 W. The range of operating frequencies defined by PMA is between 277 and 357 kHz. Unlike Qi and PMA, the Rezence standard is formulated by the Alliance for Wireless Power (W4WP) in early 2012 based on loosely coupled magnetic resonance, providing up to 50 W of power. It supports a much higher frequency of 6.78 MHz ± 15 kHz over a longer range (i.e., from a few centimeters to a few meters). It also provides greater spatial freedom for the charging device with respect to positioning. Subsequently, on June 1, 2015, the A4WP was merged with PMA to form the AirFuel Alliance. By now, there are two coexisting WPT standards, namely, Qi and AirFuel Alliance. Nonetheless, the lack of a unified wireless standard means that a wireless charger adhered to one standard is incompatible with a mobile device following another standard. Instead of passively waiting for the consolidation of these two standards, a proactive approach is to come up with a dual-standard wireless power transmitter supporting a wider range of frequencies, thereby enabling interoperability between the Qi and AirFuel standards. Practically speaking, the objective is to design a multicoil wireless power transmitter with *multiple* output (resonant) frequencies ranging from 100 to 300 kHz with a rated output power of 5 W per channel.

At the time of writing, a comprehensive prior art search reveals a number of existing multifrequency transmitter topologies and WPT techniques [1–11]. For instance, multifrequency WPT can be implemented by using a single-transmitter and multiple-receiver system with different resonant frequencies whose system architecture is shown in Figure 6.1 [1–3].

Figure 6.1 shows that there is only one single AC power source at the transmitter side. By tuning the source frequency of the transmitter to be the same as the resonant frequency of a particular receiver, selective power transfer across a specific wireless channel can be attained [1]. It is worth mentioning that only a single receiver can be powered by the transmitter at any point in time. Likewise, a selective power distribution method based

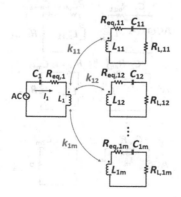

FIGURE 6.1   System architecture of a single-transmitter multiple-receiver WPT system.

on impedance matching or auxiliary circuit at the receiver is reported in [2,3]. Irrespective of the specific WPT techniques, a common drawback of this single-transmitter multiple-receiver WPT system is that it cannot achieve simultaneous power delivery to multiple receivers, which severely hampers its practical applications. To address such limitation, a multiple-transmitter multiple-receiver topology, as depicted in Figure 6.2, was presented in [4,5]. In this topology, each transmitter coil is paired up with the corresponding receiver to form a multicoil WPT system. For instance, the transmitting (primary) coil (L_1) at the transmitter side is coupled with the receiving (secondary) coil (L_{m+1}) at the receiver side.

Figure 6.2 clearly shows that the transmitting coil of each transmitter is independently powered by its own AC input source, which is tuned to the same resonant frequency of the matching network of the corresponding receiver. As a result, it is possible to produce multiple transmitting frequencies by operating the AC inputs at different frequencies. Yet, a major issue of this topology is that the number of AC power sources increases proportionally with the number of transmitting coils. It is envisaged that as the number of transmitting coils increases in a multicoil WPT system, a greater number of AC power sources are needed, which makes the overall system bulky, costly, and inefficient. In addition, since the transmitting coils are driven by independent and unsynchronized AC sources, undesirable power fluctuations across the transmitting coils will inevitably occur due to the lack of frequency synchronization. This could adversely affect the system stability. Moreover, the aforementioned topologies [1–5]

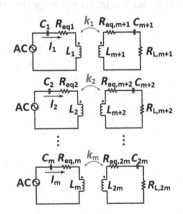

FIGURE 6.2 System architecture of a multiple-transmitter multiple-receiver WPT system.

require at least one external AC power source. Hence, they cannot be directly applied in real situations where only a DC power source is available. On the other hand, another kind of multifrequency WPT systems is reported in [6–10], which employs power inverters for powering multiple transmitting coils from one or more DC power sources. Figure 6.3 shows the system architecture of this multifrequency WPT system.

In this architecture, the power transmitter can be implemented as either one full-bridge inverter or two half-bridge inverters, which is then connected to a parallel combination of output resonant tanks to produce different resonant frequencies across the transmitting coils (L_1, L_2, ..., L_m) from a single DC input voltage (V_{in}). A double-frequency magnetically coupled resonant WPT system is realized in [7]. The transmitter is composed of two half-bridge inverters. The two bridge legs are operated at different switching frequencies, and then the output voltages of the bridge legs are subtracted from each other to generate an effective voltage for driving the transmitting resonant tank. Unfortunately, this double-frequency WPT design cannot be extended to driving multiple loads at different frequencies. A dual-frequency WPT module was also reported in [8]. It consists of a current-switching Class D inverter for driving the 6.78 MHz transmitting coil from a 12 V DC input as well as a zero-current switching (ZCS) half-bridge inverter for driving the 200 kHz transmitting coil from a 21 V DC input. Nevertheless, this approach would require multiple DC power supplies and separate inverters for driving the transmitting coils in order to enable multifrequency operation. In [9], a dual-mode WPT based on the multifrequency programmed pulse width modulation scheme was introduced. The operating principle of this modulation scheme is to adjust the switching angles of a periodic pulse train of squared waveforms in order to generate two distinct harmonics for dual-band power delivery.

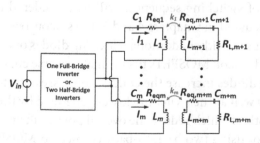

FIGURE 6.3 System architecture of a multifrequency WPT system with a single DC power source.

However, the resulting transmission frequency is not very precise because the selected higher order harmonics (e.g., 203 kHz) can never be the same as the target transmission frequency (e.g. 205 kHz). Besides, it requires complex offline computation for solving a large set of objective equations in order to derive the desired output spectrum. This is virtually impractical for real applications. Evidently, the hardware prototypes of the wireless power transmitter in [7–11] are capable of producing only two distinct transmitting frequencies simultaneously for delivering power wirelessly to two receivers. It would be rather difficult, if not impossible, to extend this prior art in such a way that they can be deployed in multifrequency WPT system with more than two receiving devices.

In light of the limitations of the prior art, a gallium nitride (GaN)-based SIMO inverter capable of producing multiple sinusoidal-like outputs with *distinct* frequencies is proposed in [12]. The concept of a SIMO inverter is formally investigated in Chapter 4. An example of this inverter known as a boost-only SIMO inverter is also introduced. A new variant of the SIMO inverter known as the buck-boost SIMO inverter with increased output power rating is also presented in Chapter 5. These two predecessors represent the *first* generation of SIMO inverter, which supports only single-frequency operation, i.e., all AC outputs resonate at the same frequency. To carry this a step further, it is natural to explore the possibility of using a SIMO inverter to generate sinusoidal-like outputs with *different* frequencies. In brief, two conditions need to be satisfied in order for a SIMO inverter to generate distinct output frequencies. First, different values of the resonant inductor and resonant capacitor must be used across the output resonant tanks to generate different resonant frequencies. Second, new switching sequences of the power switches have to be formulated in order to attain stable and sustained oscillations at each of the resonant circuits. Various types of switching sequences will be considered throughout the chapter. As discussed in Chapters 4 and 5, the silicon-based SIMO inverters require the use of either additional blocking diodes or a pair of back-to-back connected silicon MOSFETs to eliminate reverse conduction. But the extra blocking diodes increase the conduction losses due to their forward voltage drop as well as the switching losses generated by the reverse recovery current, both of which reduce the overall power efficiency. The alternative approach of using two back-to-back connected MOSFETs with their gate terminals tied together also suffers from higher conduction losses (compared to the use of a single MOSFET) due to an increase in the total on-resistance. The switching losses also increases due to an increase in MOSFET parasitic capacitances. Compared with its silicon counterpart, a GaN enhancement-mode high-electron-mobility transistor, abbreviated as GaN E-HEMT (or GaN transistor), appears to be a better alternative because it does *not* possess an intrinsic body diode. Even though the GaN transistor has no internal body diode, we simply cannot assume that it is completely switched OFF when the gate-to-source voltage (V_{GS}) is below a certain threshold. Figure 6.4 shows the relationship between I_{DS} and V_{DS} of a typical GaN-based E-HEMT [13].

From Figure 6.4, it can be observed that during its OFF state, a negative drain current (i.e., source-to-drain current) can still flow across the transistor whenever the drain-to-source voltage (V_{DS}) is below a certain negative threshold value. Yet, by applying a more negative value of V_{GS}, the drain current can be forced to zero even with a more negative V_{DS}. In other words, a more negative V_{GS} relaxes the V_{DS} bias condition required to eliminate reverse conduction of the GaN transistor. Indeed, the GaN reverse I/V relationship, as exemplified in Figure 6.4, has important implications for the gate driver circuitry. As far as the output switch (S_{out1}, S_{out2} or S_{out3}) of the SITO inverter is concerned, the source terminal of the GaN FET is tied to the AC output of the inverter. During a certain period of time within a switching cycle, the source voltage of the GaN transistor will become higher than the drain voltage, which results in a negative V_{DS}

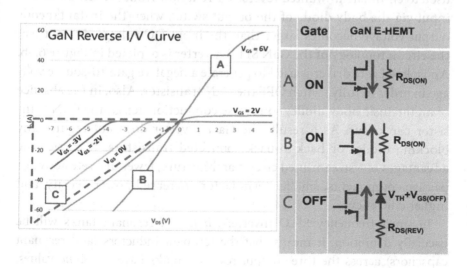

FIGURE 6.4   Relationship of I_{DS} versus V_{DS} of a typical GaN-based E-HEMT [13].

value. Thus, it is absolutely essential to ensure that the chosen isolated gate driver can provide a strong positive gate drive voltage to turn on the GaN device as well as a sufficiently negative gate drive voltage to turn it off without any reverse conduction. As a consequence, no additional blocking diodes or back-to-back-connected GaN FETs are needed in a GaN-based SIMO inverter. This leads to a simplified circuit structure, low component count, small form factor, and higher efficiency.

6.2 CIRCUIT TOPOLOGY

For the sake of comparison, the circuit topology of the silicon-based SITO boost inverter and that of the GaN-based SITO boost inverter are shown in Figure 6.5a and b, respectively.

The silicon-based inverter consists of one main inductor (L), four silicon power MOSFETs (S_{main}, S_{out1}, S_{out2}, S_{out3}), four blocking diodes (D_{main}, D_{out1}, D_{out2}, D_{out3}), three identical resonant tanks, and three independent loads. For proper circuit operation of this boost inverter, a blocking diode is connected in series with the silicon MOSFET to eliminate unintended reverse conduction. Specifically, a blocking diode D_{main} is connected in series with the main switch (S_{main}) to prevent an unintended backflow of current from ground to the negative AC output via the internal body diode of S_{main} (when S_{main} is in the OFF state). By the same token, blocking diodes at the output branches (D_{o1}, D_{o2}, D_{o3}) are used to eliminate unwanted reverse current flow from the AC to the DC input via the body diode of the output switch when the instantaneous output voltage becomes larger than the input DC voltage. In contrast, the circuit topology of the GaN SITO inverter is depicted in Figure 6.5b. An isolated gate driver is used to generate a negative gate-to-source voltage in order to fully switch OFF the GaN transistor. Also, in the absence of an intrinsic body diode, no reverse conduction across the GaN transistor can occur. As a result, the GaN inverter does *not* require any blocking diodes or back-to-back-connected transistors. It carries the advantages of a streamlined circuit architecture, fewer discrete active or passive components, smaller form factor, increased power density, and higher efficiency.

In single-frequency SITO inverters, *balanced* resonant tanks will be used. By definition, it means that the resonant inductors (and resonant capacitors) across the three output resonant tanks have the *same* values, i.e., $L_{o1} = L_{o2} = L_{o3}$ and $C_{o1} = C_{o2} = C_{o3}$. This implies that the resonant

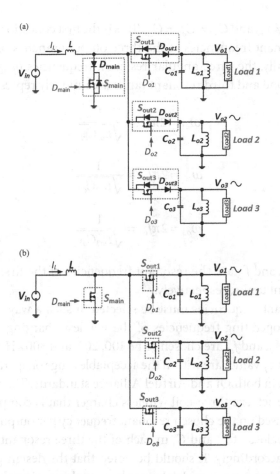

FIGURE 6.5 (a) Circuit topology of the silicon-based SITO inverter; (b) circuit topology of the GaN-based SITO inverter.

frequencies at the three output branches of the inverter are completely identical. The resonant angular frequency of each resonant tank can be expressed as

$$\omega_o = 2\pi f_o = \frac{1}{\sqrt{L_{oi}C_{oi}}}$$ (6.1)

where f_o is the resonant frequency, and i is the output index ($i = 1, 2, 3$).

In multifrequency SITO inverters, *unbalanced* resonant tanks will be used instead. Basically, it means that the resonant inductors (and resonant capacitors) across the three output resonant tanks have *different* values, i.e., $L_{o1} \neq L_{o2} \neq L_{o3}$ and $C_{o1} \neq C_{o2} \neq C_{o3}$. This is the first condition for achieving distinct resonant frequencies at the three output branches of the inverter. Mathematically, the three angular resonant frequencies (ω_{o1}, ω_{o2}, ω_{o3}) for the first, second and third resonant tanks are uniquely represented as

$$\omega_{o1} = 2\pi f_{o1} = \frac{1}{\sqrt{L_{o1}C_{o1}}} \tag{6.2a}$$

$$\omega_{o2} = 2\pi f_{o2} = \frac{1}{\sqrt{L_{o2}C_{o2}}} \tag{6.2b}$$

$$\omega_{o3} = 2\pi f_{o3} = \frac{1}{\sqrt{L_{o3}C_{o3}}} \tag{6.2c}$$

where f_{o1}, f_{o2}, and f_{o3} are the resonant frequencies of the first, second, and third resonant tanks, respectively.

The resonant frequencies must be selected in such a way that they conform to the operating frequencies of the wireless charging standard. For instance, f_{o1}, f_{o2}, and f_{o3} are chosen to be 100, 200, and 300 kHz, respectively. These frequency values are within the acceptable range of operating frequencies specified in both Qi and Airfuel Alliance standards. The SITO inverter can therefore act as a three-coil wireless charger that is compatible to both standards. Based on the chosen resonant frequency per output channel, the appropriate values of L_o and C_o in each of the three resonant tanks can be determined accordingly. It should be noted that the design parameters of the resonant circuit are completely unrelated to the semiconductor technology used in the power devices. Hence, the same resonant circuit design can be applied to both the silicon-based and GaN-based SIMO inverters. In general, depending on the actual parameter values of the resonant circuit, both types of inverters are equally capable of producing the same or different output frequencies across their outputs. But, since this chapter talks about SIMO inverter using GaN power transistors, we will show how multiple output frequencies can be achieved in this particular type of SIMO inverter.

6.3 SWITCHING SEQUENCE

It is crucial to note that the use of unbalanced LC tank circuits is a necessary but not a sufficient condition for producing multiple sinusoidal outputs of *different* frequencies. As a matter of fact, the switching sequence of the SIMO inverter also plays a significant role in providing a continuous and steady flow of energy from the main inductor, acting like a constant current source, into each of the output tank circuits. Intuitively, the output tank circuit resonating at a higher frequency (e.g. 300 kHz) would require more frequent energy injections than that resonating at a lower frequency (e.g., 100 kHz) so as to maintain sustained oscillation with constant peak amplitude. This implies that the switching sequence of the original SIMO inverters with the same output frequencies, which results in a uniform distribution of the stored energy from the main inductor across the outputs, is no longer suitable for the SIMO inverter with different output frequencies. Thus, this motivates us to derive an optimal and robust switching sequence to efficaciously perform *uneven* distribution of energy across the three nonidentical resonant tanks. In the following subsections, various kinds of switching sequences for SIMO inverter with multiple output frequecies will be presented.

6.3.1 Type I Switching Scheme

A simplistic approach of producing multiple output frequencies in a SITO inverter is to divide a full (resonant) cycle equally into three phases, of which only one output is enabled at any time instance. This is known as Type I switching scheme. Figure 6.6 shows the ideal waveforms of the three output enable signals, the gate drive signals of all power switches, and the main inductor current.

For illustration purpose, the operating (resonating) frequencies of the first, second, and third outputs of the SITO inverter are chosen to be 100, 200, and 300 kHz, respectively. The three-phase partitioning is done by the output enable signals with a slower frequency of 10 kHz. This switching scheme is characterized by the fact that the operations of each output are performed only when the corresponding output phase is enabled. Therefore, the three individual outputs are fully decoupled from one another in the time domain. In each phase, either the main switch (S_{main}) or one of the output switches (S_{out1}, S_{out2}, or S_{out3}) is flipped at the corresponding operating frequencies in a time-interleaving manner. For example, during the first phase, S_{main} and S_{out1} of the first output are turned ON (or OFF) at 100 kHz. This ensures fully independent operations among the three output channels without any cross-interference. Nevertheless, an obvious disadvantage of this simplistic switching scheme is that the output resonant tank, particularly the one with the highest frequency, will have to wait for a relatively long period of time before it can be replenished at the next phase pertaining to that output. This will result in decaying oscillations of the resonant tank within a complete cycle due to damping. This is manifested as a slow decaying of the amplitude envelope of the sinusoidal-like output voltage (assuming a weakly damped oscillator). It is conceived that as the number of outputs increases, the droop in the output voltage will become more pronounced. Hence, it is essential to come up with a better switching scheme that aims at minimizing the self-oscillation time interval of each output channel.

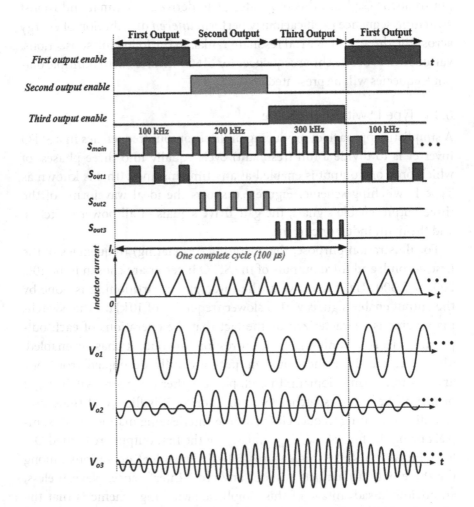

FIGURE 6.6   Ideal waveforms of the key signals of SITO inverter with Type I switching scheme.

6.3.2 Type II Switching Scheme

A main objective of the Type II switching scheme is to reduce the self-oscillation time period of each output channel in such a way that the SIMO inverter can generate smooth sinusoidal-like outputs with constant amplitudes and minimum distortion. The key is to significantly reduce the period of one full cycle. To ensure a continuous and steady supply of energy to all outputs oscillating at different frequencies, the switching frequency of the main switch (S_{main}) must be a function of *all* the resonant frequencies. This switching scheme requires the use of a carrier signal whose frequency is equal to the number of outputs in the SIMO inverter multiplied by the least common multiple (LCM) of all the resonant (output) frequencies. The switching frequency (f_{sw}) of S_{main} is the same as the carrier frequency (f_c). In general, f_{sw} can be expressed as

$$f_{sw} = f_c = n \times LCM\left(f_{o1}, f_{o2}, ... f_{on}\right) \qquad (6.3)$$

where n is the total number of output channels in a SIMO inverter, and f_{o1}, $f_{o2}, ..., f_{on}$ are the resonant frequencies.

A single-inductor dual-output (SIDO) inverter is used as an example to illustrate the Type II switching scheme. The frequencies of the two outputs are chosen to be 100 and 200 kHz. From equation (6.3), since $n = 2$ and LCM (100, 200 kHz) = 200 kHz, the switching frequency of S_{main} (f_{sw}) is therefore determined to be 400 kHz. Figure 6.7 illustrates the Type II switching sequence of the SIDO inverter. S_{out1} denotes the gate drive signal of the output switch for the *first* output channel at 100 kHz, while S_{out2} denotes the gate drive signal of the output switch for the *second* output channel at 200 kHz. In one complete cycle, S_{out2} needs to be switched *twice* as fast as S_{out1} because the resonant frequency of the second output is two times that of the first output. Because S_{main} switches at a higher frequency of 400 kHz, it takes only 10 μs to complete a full cycle in Type II versus 100 μs in Type I. The Type II method enables the two output resonators to be recharged at exactly the same rate as the resonant frequency, which leads to sustained sinusoidal oscillations of constant frequency and peak amplitude.

Even though this switching scheme works well for an SIDO inverter, it has a major issue when it is extended to a SIMO inverter with more than two outputs. For an SITO inverter with three distinct output frequencies of 100, 200, and 300 kHz, their LCM is 600 kHz, which means that f_{sw} is 1800 kHz. As can be seen in Figure 6.8, a 50% on-time duty ratio of S_{main}

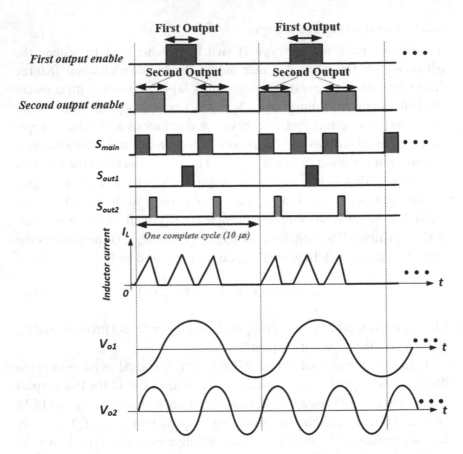

FIGURE 6.7 Ideal waveforms of the key signals of SIDO inverter with Type II switching scheme.

is equivalent to a very modest duty ratio of around 2.78% for the 100 kHz output during one complete cycle. This certainly restricts the maximum amount of energy that can be delivered to the resonant tank. In addition, the Type II scheme inevitably creates an excessive number of unused switching cycles (or idle phases) during which all power switches are OFF. Such an underutilization of switching cycles results in a relatively short charging time for the main inductor, which implies that a very limited amount of energy can be delivered from the main inductor to the output resonant tanks. This would severely limit the maximum achievable output power per channel. Moreover, equation (6.3) indicates that the switching (carrier) frequency is directly proportional to the total number of outputs. As the number of outputs of the SIMO inverter increases, the switching

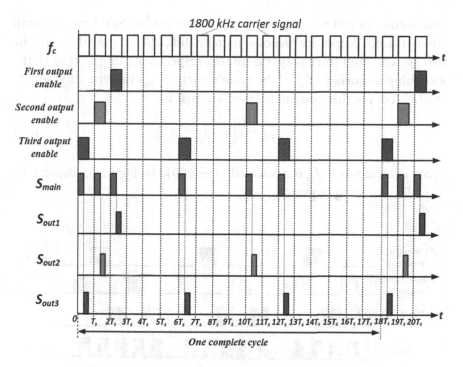

FIGURE 6.8 Ideal waveforms of the key signals of SITO inverter with Type II switching scheme.

frequency will become so high that the inverter will incur higher switching losses, thereby lowering the overall efficiency. Noise can also be a problem with higher frequencies. Jitter noise becomes a larger percentage of the switching pulse especially when the on-time duty ratio is small. Hence, the use of a very high switching frequency in SIMO inverter may not be too practical for real WPT applications.

6.3.3 Type III Switching Scheme

To address the limitations of Type I and Type II, the Type III switching scheme is introduced, which aims at satisfying the two seemingly conflicting requirements: (i) provide a continuous energy supply for the resonant tanks to produce clean sinusoidal output voltages with nearly constant amplitude and (ii) allow a sufficiently long charging time for each resonant tank to meet the rated output power. A key feature of the Type III scheme is the use of the "pulse-skipping" technique. As the name suggests, the energy transfer from the main inductor to the output of the *highest* frequency is intermittently disabled at certain switching cycles. For ease of discussion, we will revisit the previous example of an SITO inverter with three distinct output frequencies, namely, 100, 200, and 300 kHz for the first, second, and third output channels, respectively. Unlike Type II, the switching frequency (f_{sw}) of the main switch (S_{main}) in Type III is simply the LCM of the three output frequencies, that is,

$$f_{sw} = LCM\left(f_{o1}, f_{o2}, \ldots f_{on}\right) \tag{6.4}$$

From equation (6.4), f_{sw} is calculated to be 600 kHz. Figure 6.9 depicts the Type III switching sequence.

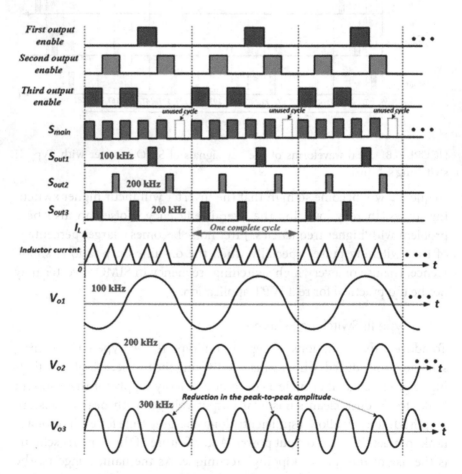

FIGURE 6.9 Ideal waveforms of the key signals of SITO inverter with Type III switching scheme.

Basically, a full cycle is evenly divided into six subcycles. Ideally, the main switch (S_{main}) should be enabled *six* times within a full cycle. Likewise, the first output switch (S_{out1}) is enabled *once*, the second output switch (S_{out2}) is enabled *twice*, and the third output (S_{out3}) is enabled *thrice* during a full cycle. The switching order of the output switches is also important. In principle, each of them should be enabled periodically at certain fixed intervals to attain a sustained energy supply to each output. Unfortunately, this ideal switching pattern does lead to a time conflict between S_{out2} and S_{out3} as both of them need to be switched on at the fifth subcycle. To resolve this conflict, a pulse-skipping method is employed which deliberately disables S_{out3} at every fifth subcycle and allows only S_{out2} to be switched on. What this means is that S_{out3} will effectively be switched on *only twice* in every full cycle. In addition, the sixth (last) subcycle is also skipped as all switches remain in their OFF state. Thus, S_{main} would effectively be switched on *five* times in a complete cycle. Even though this pulse-skipping method removes the time conflict between S_{out2} and S_{out3}, it inevitably results in unequal peak amplitudes of the third output voltage. This is mainly attributed to the lack of regular energy replenishment to maintain sustained oscillations in the third resonant tank due to "pulse-skipping" every fifth subcycle. As we shall see later, this phenomenon is also experimentally verified.

6.3.4 Modified Type III Switching Scheme

Conventionally, the operating principles of SIMO DC-DC converters underpinning the distribution of energy from a single power inductor to multiple outputs can be classified into two types, namely, *multiple* energizing phases and *single* energizing phases [12,13], which are discussed in Chapter 2. In simple terms, the former means that the main inductor is charged and discharged *separately* for each output, whereas the latter means that the main inductor is charged *only once* per switching cycle, and then its energy is released (discharged) to each output sequentially. Either of the two switching methods can be used to control the charging and discharging of the main inductor in a SIMO-based converter. The switching schemes of the SIMO DC-AC inverters we have seen so far are primarily based on multiple energizing phases. To tackle the problem of the missing fifth cycle for the third output in the Type III scheme, we can apply the method of single energizing phases specifically for the fifth cycle. In the modified Type III switching scheme, as depicted in Figure 6.10, the

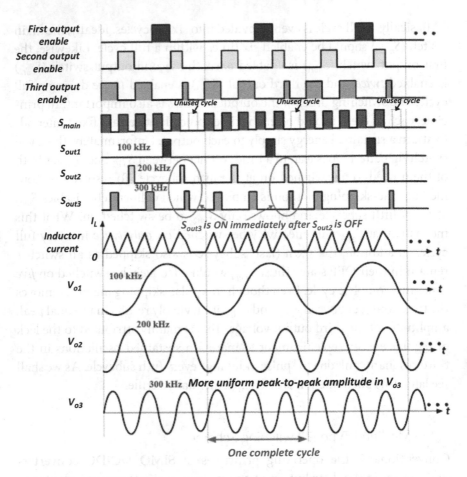

FIGURE 6.10 Ideal waveforms of the key signals of SITO inverter with modified Type III switching scheme.

third output switch (S_{out3}) is closed immediately after the second output switch (S_{out2}) is opened in the fifth switching cycle.

Unlike its predecessor, pulse-skipping is *not* required for the third output in the modified Type III scheme. S_{out3} switches at a frequency of 300 kHz, which is the same as the resonant frequency. Basically, the energy stored in the main inductor is successively transferred to the second output followed by the third output in the fifth cycle. At the end of this cycle, the main inductor is fully discharged, and the inductor current returns to zero. It is interesting to note that this modified Type III method is actually a *hybrid* switching method, which uses *both* types of switching methods, i.e., single energizing cycle (for the fifth cycle) and multiple energizing cycles (for the remaining cycles). This hybrid switching scheme enables a smoother and more uniform supply of energy from the main inductor to the third output. Thus, it reduces the droop in the peak envelope of the third output voltage. The corresponding experimental results are presented in Section 6.5.

6.4 THEORETICAL ANALYSIS

6.4.1 Proof of Sinusoidal Oscillation in Type III Switching Scheme

Before delving into the mathematical proof, it is necessary to take a closer look at the switching states and the corresponding equivalent circuits of the SITO inverter under the Type III switching scheme. Since the SITO inverter operates in DCM, each switching cycle is made up of three distinct modes of operation. The first mode of operation (or first subinterval) is the charging phase in which the main inductor is charged up while the outputs are self-oscillating. The second mode (or second subinterval) is the discharging phase in which the main inductor releases its storage energy to the output. Finally, the third mode (or third subinterval) is known as idle phase in which nothing much happens, except that the outputs are self-oscillating. In this mode, the main switch and all output switches are opened, while the main inductor current remains at zero value.

In general, the Type III switching sequence comprises a total of 18 switching states per one complete cycle. Each state can be modeled by an equivalent circuit of the SITO inverter, as shown in Figure 6.11. By default, each state is uniquely represented by two integers (x, y), where x is the switching cycle number, and y is the mode of operation in DCM. The value of x ranges from 1 to 6 as there are *six* switching cycles in a full cycle. The value of y ranges from 1 to 3 because of *three* modes of operation in DCM. There are three states in each cycle, namely, $(x, 1)$, $(x, 2)$, and $(x, 3)$. For instance, the state denoted by (1–2) corresponds to the *first* cycle and the *second* mode of operation. In this particular state, the main inductor releases its energy to the *third* output, which is indicated by the corresponding equivalent circuit in Figure 6.11. States (1–1), (2–1), (3–1), (4–1), and (5–1) share the same equivalent circuit because each switching cycle always begins with the charging of the main inductor. Since all output switches are opened, the self-oscillating output resonant tanks are completely separated from the main inductor. The equivalent circuits for states (1–2), (2–2), (3–2), (4–2), and (5–2) are determined by the switching order of the outputs. As shown in Figure 6.9, the first output is enabled in the fourth cycle, the second output is enabled in the second and fifth cycles, and the third output is enabled in the first and third cycles. States (1–3), (2–3), (3–3), (4–3), and (5–3) have

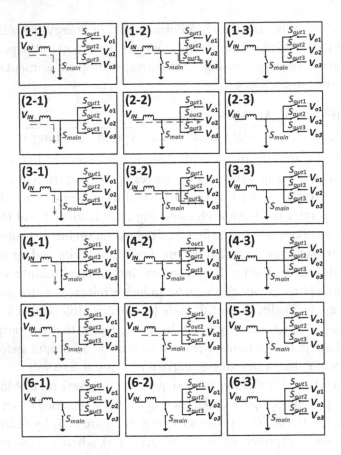

FIGURE 6.11 Switching states and equivalent circuits of the SITO inverter with Type III switching scheme.

the same equivalent circuit since the third subinterval of every switching cycle must be an idle phase. Last but not least, states (6–1), (6–2), and (6–3) in the sixth (last) switching cycle also share a common circuit in that the three outputs are disabled, and all switches remain opened. The output resonant tanks are in self-oscillating mode. Since no switching actions occur in this cycle, it can be treated as an idle cycle.

Since the SITO inverter employs multiple-energizing phases between the first and second outputs, these two output channels are safely decoupled from each other in time without any cross-interference. Besides, the output switches pertaining to these two outputs are enabled periodically without any disruptions. The mathematical proof of sinusoidal oscillations from Chapter 3 continues to hold for the first two channels, and so, it will not be repeated here. However, in the Type III switching method, the fact that the third output is disabled at every fifth switching cycle distinguishes it from the first two outputs. It is therefore necessary to perform theoretical analysis exclusively for the third output and more importantly, to ascertain that sinusoidal oscillations can also be achieved at the third output.

In addition to the three normal modes of operation in DCM, a new mode (i.e., Mode 4) is introduced for the third output operating at 300 kHz. Figure 6.12 highlights the time interval of Mode 4, which starts at time t_6 and ends at time T, where T is a complete (resonant) cycle. Since a complete (resonant) cycle is made up of six switching cycles, we can therefore write: $T = 6T_s$, where T_s is the switching period. For the sake of discussion, this figure is specifically used to illustrate the switching order of the main switch (S_{main}) corresponding to the third output as well as the third output switch (S_{out3}).

Mode 4 is used to take into account the disruption in the switching sequence of S_{main} and S_{out3} due to the use of pulse skipping. In essence, it can be considered as a pseudo-switching cycle in which S_{main} and S_{out3} are switched OFF. In the following subsection, a complete mathematical proof of sinusoidal oscillations for the third output in each mode of operation will be presented. To simplify the ensuing analysis, the SITO inverter with *only* the third output enabled is effectively modeled as a single-inductor single-output (SISO) inverter.

A. Mode 1–Proof of Sinusoidal Oscillation

Figure 6.12 shows the circuit diagram of an ideal SISO boost inverter operating in Mode 1. In this mode, S_{main} is switched ON, whereas S_{out3} is switched OFF. This corresponds to the *first* subinterval in DCM during which the main inductor (L) is charged up

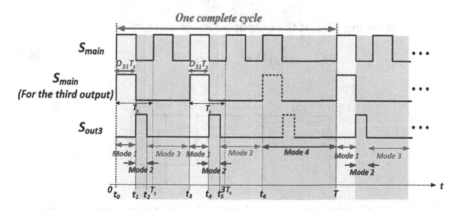

FIGURE 6.12 Modes of operation exclusively for the third output.

by DC power supply. Since S_{out3} is OFF in this mode, the LC resonator is completely isolated from the input stage, and therefore, it self-oscillates at the resonant frequency. Yet, as time goes by, the storage energy in the resonator will be dissipated by the parasitic resistances as well as the resistive loads R_L. This causes the decay of the peak envelope of the third output voltage. Figure 6.13 shows the equivalent circuit diagram of the power stage of the SISO inverter operating in Mode 1. It is interesting to note that R_L is being modeled as a variable resistor. In a typical two-coil WPT system, the transmitting coil is closely coupled with the loaded receiving coil. As a first-order approximation, this can be modeled by a variable resistor at the primary (transmitter) side whose value will vary depending on the actual distance between the transmitting and receiving coils.

A closed circuit is formed when S_{main} is turned ON. The inductor current will flow from the input voltage (V_{in}) to S_{main} and ground via the main inductor (L). A voltage will develop across L which can be expressed as

$$V_{in} = L\frac{di_L}{dt} = L\frac{I_{L,peak}}{D_{31}T_s} \tag{6.5}$$

where i_L is the inductor current, $I_{L,\,peak}$ is the peak value of the inductor current, D_{31} is the on-time duty ratio of S_{main} pertaining to the third output, and T_s is the switching period.

By rearranging the terms in equation (6.5), $I_{L,\,peak}$ can be written as

$$I_{L,peak} = \frac{V_{in}D_{31}T_s}{L} \tag{6.6}$$

By applying KCL at the output node, we can write

$$i_{Co}(t)+i_{Lo}(t)+i_R = 0 \tag{6.7}$$

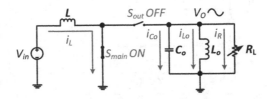

FIGURE 6.13   Circuit diagram of an ideal SISO boost inverter operating in Mode 1.

Then, by using Laplace transform, equation (6.7) can be transformed to the s-domain as follows.

$$C_o\left[sV_o(s)-V_{oi}\right]+\frac{V_o(s)}{L_oS}+\frac{I_{Loi}}{S}+\frac{V_o(s)}{R_L}=0 \tag{6.8}$$

where V_{oi} and I_{Loi} are the initial values of the output voltage and the resonant inductor current, respectively. That is, $V_{oi}=v_o(t_0)$ and $I_{Loi}=i_{Lo}(t_0)$, where t_0 is the start time of Mode 1 [see Figure 6.12].

By rearranging the terms in equation (6.8), the output voltage can therefore be expressed as

$$V_o(s)=\left[\frac{s}{\left(s+\dfrac{1}{2R_LC_o}\right)^2+\dfrac{1}{L_oC_o}-\dfrac{1}{4R_L^2C_o^2}}\right]V_{oi}$$

$$-\left[\frac{1}{\left(s+\dfrac{1}{2R_LC_o}\right)^2+\dfrac{1}{L_oC_o}-\dfrac{1}{4R_L^2C_o^2}}\right]\frac{I_{Loi}}{C_o} \tag{6.9}$$

Since $\dfrac{1}{L_oC_o}>>\dfrac{1}{4R_L^2C_o^2}$, equation (6.9) can be reduced to the following form.

$$V_o(s)=\left[\frac{s}{\left(s+\dfrac{1}{2R_LC_o}\right)^2+\dfrac{1}{L_oC_o}}\right]V_{oi}-\left[\frac{1}{\left(s+\dfrac{1}{2R_LC_o}\right)^2+\dfrac{1}{L_oC_o}}\right]\frac{I_{Loi}}{C_o} \tag{6.10}$$

By applying inverse Laplace transform to equation (6.10), the output voltage in the time domain can be obtained as follows.

$$v_o(t)=V_{oi}e^{-\frac{1}{2R_LC_o}t}\cos\omega_0t-\frac{I_{Loi}}{C_o\omega_0}e^{-\frac{1}{2R_LC_o}t}\sin\omega_0t-\frac{V_{oi}}{2R_LC_o\omega_0}e^{-\frac{1}{2R_LC_o}t}\sin\omega_0t$$

$$\tag{6.11}$$

where $\omega_0=\dfrac{1}{\sqrt{L_oC_o}}$ and $0\le t\le t_1$, $t_1=D_{31}T_s$.

Equation (6.11) shows that the output voltage is sinusoidal in nature and its fundamental frequency is the same as the resonant frequency of the LC tank circuit. As a sanity check, by substituting $t = 0$ into equation (6.11), $v_o(0) = V_{oi}$, which confirms that $v_o(0)$ is indeed the initial value of the output voltage in Mode 1. In particular, the exponential decay functions in equation (6.11) imply that in the absence of any input power injection, the peak amplitude of the output voltage will decrease exponentially.

B. Mode 2–Proof of Sinusoidal Oscillation

Figure 6.14 shows the circuit diagram of an ideal SISO boost inverter operating in Mode 2. S_{main} is switched OFF, and S_{out} is switched ON. This corresponds to the *second* subinterval in DCM. Since S_{out} is closed, the resonant tank is connected to the main inductor (L), which allows the energy stored in L to be transferred to the resonant circuit and the output load. As L continues to discharge, its current will drop until it finally reaches zero (which means L is fully discharged). It should be noted that Mode 2 is the only mode per switching cycle in which the energy in the resonant circuit is replenished for achieving sustained oscillations in the output voltage with constant frequency and peak amplitude.

By applying KVL in the closed-loop circuit in Figure 6.14, we have

$$V_{in} - V_o(t) = L\frac{di_L(t)}{dt} \tag{6.12}$$

Equation (6.12) can be transformed from the time domain to the frequency domain (s-domain) by using Laplace transform. Hence, equation (6.12) becomes

$$\frac{V_{in}}{s} - V_o(s) = sLI_L(s) - LI_{Li} \tag{6.13}$$

where I_{Li} is the initial value of the inductor current, i.e., $I_{Li} = i_L(t_1)$, and t_1 is the start time of Mode 2.

By applying KCL at the output node, we have

$$i_{Co}(t) + i_{Lo}(t) + i_R = i_L(t) \tag{6.14}$$

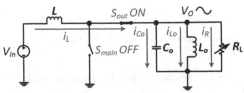

FIGURE 6.14 Circuit diagram of an ideal SISO inverter operating in Mode 2.

Likewise, equation (6.14) can be transformed from the time domain to the s-domain as follows.

$$C_o\left[sV_o(s)-V_{oi}\right]+\frac{V_o(s)}{L_o s}+\frac{I_{Loi}}{s}+\frac{V_o(s)}{R_L}=I_L(s) \tag{6.15}$$

where V_{oi} and I_{Loi} are the initial values of the output voltage and the resonant inductor current, i.e., $V_{oi}=v_o(t_1)$, $I_{Loi}=i_{Lo}(t_1)$, and t_1 is the start time of Mode 2.

Notice that the initial value of the inductor current (I_{Loi}) is identical to the peak value of the inductor current ($I_{L,peak}$), which is given by equation (6.6). By substituting equations (6.6) and (6.15) into (6.13), we have

$$\frac{V_{in}}{s}=sL\left[sC_oV_o(s)-C_oV_{oi}+\frac{V_o(s)}{L_o s}+\frac{I_{Loi}}{s}+\frac{V_o(s)}{R_L}\right]$$

$$+V_o(s)-V_{in}D_{31}T_s \tag{6.16}$$

By rearranging the terms in equation (6.16), $V_o(s)$ can be expressed as

$$V_o(s)=\frac{1}{s^3+\dfrac{1}{R_LC_o}s^2+\dfrac{L+L_o}{L_oLC_o}s}\frac{V_{in}}{LC_o}+\frac{s}{s^2+\dfrac{1}{R_LC_o}s+\dfrac{L+L_o}{L_oLC_o}}V_{oi}$$

$$+\frac{1}{s^2+\dfrac{1}{R_LC_o}s+\dfrac{L+L_o}{L_oLC_o}}\frac{V_{in}D_{31}T_s}{LC_o}-\frac{1}{s^2+\dfrac{1}{R_LC_o}s+\dfrac{L+L_o}{L_oLC_o}}\frac{I_{Loi}}{C_o} \tag{6.17}$$

Let $\omega_1=\sqrt{\dfrac{1}{L_oC_o}+\dfrac{1}{LC_o}}$ and apply inverse Laplace transform to equation (6.17), a general expression of the output voltage in the time domain can be obtained as

$$v_o(t)=\frac{L_o}{L_o+L}V_{in}-\frac{L_o}{L_o+L}V_{in}e^{-\frac{1}{2R_LC_o}t}\cos\omega_1 t-\frac{L_o}{L_o+L}\left(\frac{V_{in}}{2R_LC_o\omega_1}\right)e^{-\frac{1}{2R_LC_o}t}\sin\omega_1 t$$

$$+\frac{V_{in}D_{31}T_s}{LC_o}\left(\frac{1}{\omega_1}\right)e^{-\frac{1}{2R_LC_o}t}\sin\omega_1 t+V_{oi}e^{-\frac{1}{2R_LC_o}t}\cos\omega_1 t$$

$$-\frac{V_{oi}}{2R_LC_o\omega_1}e^{-\frac{1}{2R_LC_o}t}\sin\omega_1 t-\frac{I_{Loi}}{C_o\omega_1}e^{-\frac{1}{2R_LC_o}t}\sin\omega_1 t \tag{6.18}$$

where $t_1\le t\le t_2$, $t_2=t_1+D_{32}T_s$, and D_{32} is the duty ratio of S_{out} corresponding to the third output.

Assume $L \gg L_o$. The resonant frequency in Mode 2 will be the same as that in Mode 1, i.e., $\omega_1 = \omega_0$. In the event that this assumption does not hold, the harmonic distortion of the sinusoidal-like output voltage can still be minimized by using a smaller time interval for Mode 2. On the other hand, it can be seen from Figure 6.14 that a constant DC input voltage (V_{in}) is effectively applied across the series-connected L and L_o. By treating this as a voltage divider, the DC voltage across L_o is basically the DC component (or DC offset) of the output voltage. This DC component is mathematically represented by the first term in equation (6.18), which has a constant value. Also, it is the only term in equation (6.18) that is frequency-independent. In fact, this term can be physically interpreted as the DC power source that is responsible for injecting power into the resonant tank.

C. Mode 3–Proof of Sinusoidal Oscillation

In this idle mode, both S_{main} and S_{out} are in their OFF positions. Figure 6.15 shows the circuit diagram of an ideal SISO boost inverter operating in Mode 3. A careful observation reveals that this circuit is no different from that of Mode 1. This is because in either of these modes, the resonant tank is completely isolated from the input stage due to the opening of S_{out}. The resonant tank in self-resonance serves as the only energy source for the output load.

By applying KCL at the output node, we have

$$i_{Co}(t) + i_{Lo}(t) + i_R = 0 \tag{6.19}$$

In Mode 3, the inductor current (i_L) remains at zero value. Because the circuit in Mode 3 is basically the same as that in Mode 1, the output voltage can therefore be written as

$$v_o(t) = V_{oi}e^{-\frac{1}{2RC_o}t}\cos\omega_0 t - \frac{I_{Loi}}{C_o\omega_0}e^{-\frac{1}{2RC_o}t}\sin\omega_0 t$$

$$-\frac{V_{oi}}{2R_LC_o\omega_0}e^{-\frac{1}{2RC_o}t}\sin\omega_0 t \tag{6.20}$$

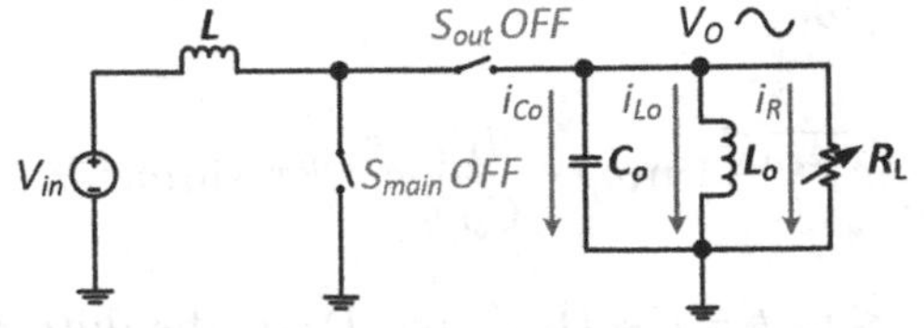

FIGURE 6.15   Circuit diagram of an ideal SISO inverter operating in Mode 3.

where V_{oi} and I_{Loi} are the initial values for the output voltage and the resonant inductor current in Mode 3, i.e., $V_{oi} = v_o(t_2)$, $I_{Loi} = i_{Lo}(t_2)$, and $t_2 \leq t \leq t_2 + D_{33}T_s$.

D. Mode 4–Proof of Sinusoidal Oscillation

Mode 4 is a new mode specifically for the third output. Figure 6.16 depicts the circuit diagram of an ideal SISO boost inverter operating in Mode 4. It shows that both S_{main} and S_{out} are in their OFF positions. A major difference between Mode 4 and the other modes is that Mode 4 literally spans two switching cycles, i.e., from $t = t_6$ to $t = T$ [see Figure 6.12]. Since the fifth switching cycle is specifically reserved for the second output and the sixth switching cycle is unused by all outputs, the third output remains disabled (i.e. S_{out3} is OFF). In other words, the LC resonator serves as the only power source for the third output during the last two cycles. Because these two cycles are skipped, the third output voltage experiences a gradual decay in its peak-to-peak amplitude. It is also worth noting that since the main inductor is *not* connected to the resonant tank at all times during Mode 4, the output voltage is purely sinusoidal with no harmonic distortion.

Since the equivalent circuit in Mode 4 is identical to that in Mode 1 or 3, the output voltage can be expressed as

$$v_o(t) = V_{oi}e^{-\frac{1}{2R_LC_o}t}\cos\omega_0 t - \frac{I_{Loi}}{C_o\omega_0}e^{-\frac{1}{2R_LC_o}t}\sin\omega_0 t$$

$$-\frac{V_{oi}}{2R_LC_o\omega_0}e^{-\frac{1}{2R_LC_o}t}\sin\omega_0 t \tag{6.21}$$

where V_{oi} and I_{Loi} are the initial values of the output voltage and resonant inductor current in Mode 4, i.e., $V_{oi} = v_o(t_6)$, $I_{Loi} = i_{Lo}(t_6)$, and $t_6 \leq t \leq T$.

6.4.2 Generalized Mathematical Description

The expressions of the output voltage for the third output channel previously derived apply only to the first full (resonant) cycle. In this section, a general expression of the output voltage in each of the four operating

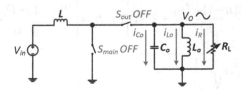

FIGURE 6.16 Circuit diagram of an ideal SISO inverter operating in Mode 4.

modes in steady-state condition is presented, which is applicable to *any* resonant (or switching) cycle.

Mode 1: For $t \in [t_0, t_1]$,

$$v_o(t) = v_o(t_0) e^{-\frac{1}{2R_L C_o}t} \cos \omega_0 t - \frac{i_{Lo}(t_0)}{C_o \omega_0} e^{-\frac{1}{2R_L C_o}t} \sin \omega_0 t$$

$$- \frac{v_o(t_0)}{2R_L C_o \omega_0} e^{-\frac{1}{2R_L C_o}t} \sin \omega_0 t$$

(6.22)

where $t_0 = \left[\frac{2}{6}(n-1)+k-1\right]T, t_1 = t_0 + D_{31}T_s, n = \left\{x \in Z \mid x = [1,2]\right\}, k \in Z^+,$

and $\omega_0 = \frac{1}{\sqrt{L_o C_o}}$.

It should be noted that n is a positive integer, which represents the nth enabled switching cycle of the third output in each complete cycle. Since the third output is enabled only twice within a full resonant period [see Figure 6.12], the value of n can either be 1 or 2. The integer k stands for the kth full cycle. Recall that there are six switching cycles within a full cycle. Hence, for the first switching cycle ($n = 1$) in the first full cycle ($k = 1$), $t_0 = 0$ and $t_1 = D_{31}T_s$.

Mode 2: For $t \in [t_1, t_2]$,

$$v_o(t) = \frac{L_o}{L_o + L} V_{in} - \frac{L_o}{L_o + L} V_{in} e^{-\frac{1}{2R_L C_o}t} \cos \omega_1 t - \frac{L_o}{L_o + L} \frac{V_{in}}{2R_L C_o \omega_1} e^{-\frac{1}{2R_L C_o}t} \sin \omega_1 t$$

$$+ \frac{V_{in} D_1 T}{L C_o} \frac{1}{\omega_1} e^{-\frac{1}{2R_L C_o}t} \sin \omega_1 t + v_o(t_1) e^{-\frac{1}{2R_L C_o}t} \cos \omega_1 t$$

(6.23)

$$- \frac{v_o(t_1)}{2R_L C_o \omega_1} e^{-\frac{1}{2R_L C_o}t} \sin \omega_1 t - \frac{i_{Lo}(t_1)}{C_o \omega_1} e^{-\frac{1}{2R_L C_o}t} \sin \omega_1 t$$

where $t_1 = \left[\frac{2(n-1)+D_{31}}{6}+k-1\right]T, \ t_2 = \left[\frac{2(n-1)+D_{31}+D_{32}}{6}+k-1\right]T,$

and $\omega_1 = \sqrt{\frac{1}{L_o C_o} + \frac{1}{L C_o}}$.

Mode 3: For $t \in [t_2, t_3]$,

$$v_o(t) = v_o(t_2)e^{-\frac{1}{2R_LC_o}t}\cos\omega_0 t - \frac{i_{Lo}(t_2)}{C_o\omega_0}e^{-\frac{1}{2R_LC_o}t}\sin\omega_0 t - \frac{v_o(t_2)}{2R_LC_o\omega_0}e^{-\frac{1}{2R_LC_o}t}\sin\omega_0 t$$

(6.24)

where $\quad t_2 = \left[\dfrac{2(n-1)+D_{31}+D_{32}}{6}+k-1\right]T, \quad t_3 = \left(\dfrac{2}{6}n+k-1\right)T, \quad$ and

$\omega_0 = \dfrac{1}{\sqrt{L_oC_o}}.$

Mode 4: For $t \in [t_6, kT]$,

$$v_o(t) = v_o(t_6)e^{-\frac{1}{2R_LC_o}t}\cos\omega_0 t - \frac{i_{Lo}(t_6)}{C_o\omega_0}e^{-\frac{1}{2R_LC_o}t}\sin\omega_0 t - \frac{v_o(t_6)}{2R_LC_o\omega_0}e^{-\frac{1}{2R_LC_o}t}\sin\omega_0 t$$

(6.25)

where $t_6 = \left(\dfrac{4}{6}+k-1\right)T$ and $\omega_0 = \dfrac{1}{\sqrt{L_oC_o}}.$

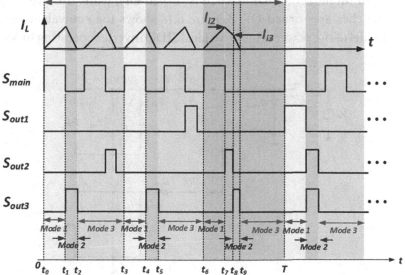

FIGURE 6.17 Ideal waveforms of the inductor current and the gate drive signals of all switches under the modified Type III switching scheme.

6.4.3 Proof of Sinusoidal Oscillation in Modified Type III Switching Scheme

Unlike its counterparts, the modified Type III is a *hybrid* scheme that combines the two different switching techniques of SIMO converters. It uses a single energizing phase in every fifth switching cycle and multiple energizing phases in the remaining switching cycles. In the fifth cycle, the energy stored in the main inductor will be transferred to the second output and then the third output in a sequential manner. In other words, the third output will be enabled immediately after the second output is disabled. Figure 6.17 shows the ideal waveforms of the inductor current along with the gate drive signals of the main switch and output switches under the modified Type III switching scheme.

Compared to the other switching cycles, the fifth cycle is unique in that *two* outputs are enabled in this cycle. Hence, both the second and third outputs need to be included in the circuit analysis. Consequently, an ideal SIDO inverter will be used to model the circuit operations of this modified Type III switching scheme. Note that only the fifth cycle is considered in the following proof of sinusoidal oscillations of the modified Type III switching scheme.

A. Mode 1–Proof of Sinusoidal Oscillation

In Mode 1, the main switch is turned ON, while the two output switches are turned OFF. Figure 6.18 shows the equivalent circuit model of the power stage of an ideal SIDO inverter operating in Mode

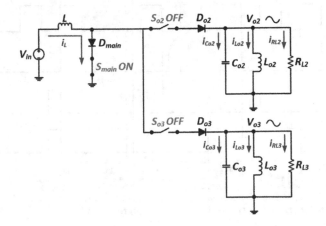

FIGURE 6.18 Equivalent circuit model of the power stage of SIDO inverter operating in Mode 1.

1. The input stage of the inverter is fully decoupled from the two AC outputs. The main inductor (L) is charged up, while the two output LC tanks self-oscillate at their respective resonating frequencies.

The circuit operation in this mode is fundamentally the same between the original Type III and modified Type III. Hence, equation (6.11) continues to hold for the second (or third) output. For the purpose of distinguishing between the two outputs, a subscript x is introduced for some state variables and the angular frequency, where $x = 2$ corresponds the second output, and $x = 3$ corresponds to the third output.

$$v_{ox}(t) = V_{oxi}e^{-\frac{1}{2R_{Lx}C_{ox}}t}\cos\omega_{0x}t - \frac{I_{Loxi}}{C_{ox}\omega_{0x}}e^{-\frac{1}{2R_{Lx}C_{ox}}t}\sin\omega_{0x}t$$

$$-\frac{V_{oxi}}{2R_{Lx}C_{ox}\omega_{0x}}e^{-\frac{1}{2R_{Lx}C_{ox}}t}\sin\omega_{0x}t \qquad (6.26)$$

where $\omega_{ox} = \dfrac{1}{\sqrt{L_{ox}C_{ox}}}$, $t_6 \leq t \leq t_7$, $t_7 = t_6 + D_1T_s$, and $x = 2, 3$.

Equation (6.26) states that the second (or third) output voltage is sinusoidal in nature whose fundamental frequency is the same as the resonant frequency of the LC tank circuit. The physical significance of the exponential decay functions in equation (6.26) is that *without* any external power injection, the sinusoidal oscillations of the LC tank will become damped over time due to internal losses.

B. Mode 2–Proof of Sinusoidal Oscillation

In Mode 2, the main switch is turned OFF, while the second and third output switches are turned ON successively, as shown in Figure 6.17. Figure 6.19a shows the equivalent circuit model of the power stage of an ideal SIDO inverter operating in Mode 2 when the second output is enabled, and the third output is disabled. Conversely, Figure 6.19b shows the equivalent circuit model of the same inverter operating in Mode 2 when the second output is disabled, and the third output is enabled.

Figure 6.19a shows that a closed loop is formed by the input voltage (V_{in}) and the second output (V_{o2}) via the main inductor (L) and the second output switch (S_{o2}). Likewise, Figure 6.19b shows that

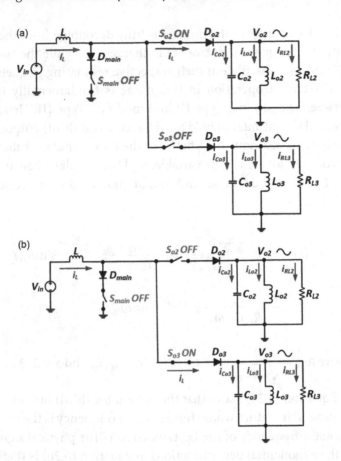

FIGURE 6.19 Equivalent circuit model of the power stage of the SIDO inverter operating in Mode 2 when (a) the second output is enabled, and the third output is disabled; (b) the second output is disabled, and the third output is enabled.

a closed loop is formed by the input voltage (V_{in}) and the third output (V_{o3}) via the main inductor (L) and the third output switch (S_{o3}). By invoking KVL in either of the two loops, we have

$$V_{in} - V_{ox}(t) = L\frac{di_L(t)}{dt} \tag{6.27}$$

By using Laplace transform, equation (6.27) can be transformed into equation (6.28) as follows.

$$\frac{V_{in}}{s} - V_{ox}(s) = sLI_L(s) - LI_{Lxi} \tag{6.28}$$

Now, by invoking KCL at either the second or third output node, we can write

$$i_{Cox}(t)+i_{Lox}(t)+i_{RLx}=i_L(t) \tag{6.29}$$

By using Laplace transform, equation (6.29) can be transformed into equation (6.30) as follows.

$$C_{ox}\left[sV_{ox}(s)-V_{oxi}\right]+\frac{V_{ox}(s)}{L_{ox}s}+\frac{I_{Loxi}}{s}+\frac{V_{ox}(s)}{R_L}=I_L(s) \tag{6.30}$$

By substituting equation (6.30) into (6.28), we have

$$\frac{V_{in}}{s}=sL\left[sC_{ox}V_{ox}(s)-C_{ox}V_{oxi}+\frac{V_{ox}(s)}{L_{ox}s}+\frac{I_{Loxi}}{s}+\frac{V_{ox}(s)}{R_L}\right]+V_{ox}(s)-LI_{Lxi} \tag{6.31}$$

Finally, by rearranging the terms, $V_{ox}(s)$ can be expressed in the following form.

$$V_{ox}(s)=\frac{1}{s^3+\frac{1}{R_LC_{ox}}s^2+\left(\frac{1}{L_{ox}C_{ox}}+\frac{1}{LC_{ox}}\right)s}\frac{V_{in}}{LC_{ox}}$$

$$+\frac{1}{s^2+\frac{1}{R_LC_{ox}}s+\left(\frac{1}{L_{ox}C_{ox}}+\frac{1}{LC_{ox}}\right)}\frac{LI_{ix}}{LC_{ox}}+\frac{s}{s^2+\frac{1}{R_LC_{ox}}s+\left(\frac{1}{L_{ox}C_{ox}}+\frac{1}{LC_{ox}}\right)}V_{oxi}$$

$$-\frac{1}{s^2+\frac{1}{R_LC_{ox}}s+\left(\frac{1}{L_{ox}C_{ox}}+\frac{1}{LC_{ox}}\right)}\frac{I_{Loxi}}{C_{ox}} \tag{6.32}$$

Let $\omega_{1x}=\sqrt{\frac{1}{L_{ox}C_{ox}}+\frac{1}{LC_{ox}}}$ and apply inverse Laplace transform in equation (6.32); the general expression of the output voltage in the time domain can be derived as follows.

$$v_{ox}(t_{otx}) = \frac{L_{ox}}{L_{ox}+L}V_{in} - \frac{L_{ox}}{L_{ox}+L}V_{in}e^{-\frac{1}{2R_LC_{ox}}t}\cos\omega_{1x}t$$

$$-\frac{L_{ox}}{L_{ox}+L}\left(\frac{V_{in}}{2R_LC_{ox}\omega_1}\right)e^{-\frac{1}{2R_LC_{ox}}t}\sin\omega_{1x}t + \frac{I_{Lxi}}{C_{ox}}\left(\frac{1}{\omega_{1x}}\right)e^{-\frac{1}{2R_LC_{ox}}t}\sin\omega_{1x}t$$

$$-\frac{V_{oxi}}{2R_LC_{ox}\omega_{1x}}e^{-\frac{1}{2R_LC_{ox}}t}\sin\omega_{1x}t + V_{oxi}e^{-\frac{1}{2R_LC_{ox}}t}\cos\omega_{1x}t$$

$$-\frac{I_{Loxi}}{C_{ox}\omega_{1x}}e^{-\frac{1}{2R_LC_{ox}}t}\sin\omega_{1x}t \tag{6.33}$$

where I_{Lxi} is the initial value of the main inductor current when either the second or third output switch is turned ON.

As far as the second output is concerned, I_{Li2} is the peak value of the main inductor current [see Figure 6.17], where $I_{Li2} = (V_{in}D_{21}T_S)/L$. The time interval of Mode 2 associated with the second output is given by $t_7 \leq t_{ot2} \leq t_8$, where $t_8 = t_7 + D_{22}T_S$, and D_{22} is the on-time duty ratio of the second output switch (S_{out2}). On the other hand, for the third output, I_{Li3} is the instantaneous value of the main inductor current when S_{out2} is turned OFF, and the third output switch (S_{out3}) is turned ON. The time interval of Mode 2 associated with the third output is given by $t_8 \leq t_{ot3} \leq t_9$, where $t_9 = t_8 + D_{23}T_S$, and D_{23} is the on-time duty ratio of S_{out3}. Equation (6.33) shows that the output voltage in Mode 2 is sinusoidal-like with a fundamental frequency of $\omega_{1x} = \sqrt{\dfrac{1}{L_{ox}C_{ox}} + \dfrac{1}{LC_{ox}}}$, where the second and third outputs are denoted by $x = 2$ and $x = 3$, respectively. It is interesting to note that equation (6.33) closely resembles equation (6.18) representing the Type III switching scheme. Notice that the assumptions in the Type III switching scheme for attaining minimum harmonic distortion remain valid for modified the Type III scheme. Hence, they will not be repeated here.

C. Mode 3–Proof of Sinusoidal Oscillation

In Mode 3, the main switch and the two outputs switches are opened. Consequently, the second and third outputs operate in self-resonance. The equivalent circuit is virtually the same as that in Mode 1. Based on equation (6.26), the expression of the second (or third) output voltage in Mode 3 can be written as

$$v_{ox}(t) = V_{oxi} e^{-\frac{1}{2R_L C_{ox}}t} \cos \omega_{ox} t - \frac{I_{Loxi}}{C_{ox} \omega_{ox}} e^{-\frac{1}{2R_L C_{ox}}t} \sin \omega_{ox} t$$

$$-\frac{V_{oxi}}{2R_L C_{ox} \omega_{ox}} e^{-\frac{1}{2R_L C_{ox}}t} \sin \omega_{ox} t$$

(6.34)

where $\omega_{ox} = \sqrt{\dfrac{1}{L_{ox} C_{ox}}}$, $t_8 \le t \le T$ for the second output, $t_9 \le t \le T$ for the third output, and $x = 2, 3$.

Recall that in the modified Type III scheme, there is no pulse-skipping for the third output. In other words, the third output is enabled three times within a full (resonant) period. Hence, mode 4 from Type III does *not* apply in this case. In short, the above theoretical derivations show that the SITO inverter using the modified Type III switching scheme can produce sinusoidal-like output voltages.

6.5 EXPERIMENTAL VERIFICATION

A series of experiments are conducted to demonstrate the effectiveness of the GaN-based SITO inverter. As a starting point, the SITO inverter is configured to operate with the same (unified) resonant frequencies across the three outputs. This is mainly to verify the basic functionality of the power stage. Afterwards, the Type III and modified Type III switching sequences will be applied to the inverter. This helps to validate that the SITO inverter can produce three different resonant frequencies across the outputs. Figure 6.20 shows a photograph of the experimental setup of a three-coil WPT system, which includes the GaN SITO inverter (acting as a three-coil wireless power transmitter), the inductive link (three pairs of transmitting-receiving coils), and three loaded receivers. Table 6.1 provides a list of part numbers of the key components used in the SITO inverter.

6.5.1 Same Output Frequencies

Table 6.2 contains the design specifications of the SITO boost-type inverter. In this experiment, the three output resonant tanks of the SITO inverter are completely identical with the same values for the resonant inductor

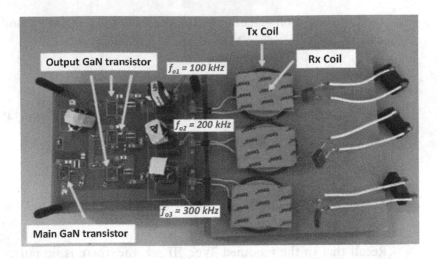

FIGURE 6.20 Experimental setup of a three-coil wireless power transfer (WPT) system.

TABLE 6.1 Component List of the Proposed Inverter

Component	Part Number
Power stage	
Main inductor (L)	Custom-made
GaN transistor	GS61004B-E01-MR
Resonant capacitor	ECW-F4364HL, ECW-F4224JB
Resonant inductor (L_{o1}, L_{o2}, L_{o3})	Custom-made
Transmitting coil	760308100110
Receiving coil	760308103211
Output load resistor (R_{L1}, R_{L2}, R_{L3})	RCH25S22R00JS06
Isolated driver	
Isolated gate driver	SI8261BAC-C-IS
Isolated transformer driver	MAX845
Transformer	TGM-010P3RL

(6.8 μH) and resonant capacitor (0.37 μF). The resonant frequency of each output is 100 kHz, which falls within the range of operation frequencies specified in the Qi standard [14]. Because the inverter has three outputs, the switching frequency of the main switch is 300 kHz, which is three times that of the output switch. A constant DC input voltage of 5 V is used throughout the experiments. Each output is connected to a 22 Ω chassis mount resistor. Balanced load condition is investigated, i.e., the three outputs have the same power rating.

Figure 6.21 shows the measured waveforms of the gate drive signals of the main switch (S_{main}) and the three output switches (S_{out1}, S_{out2}, and S_{out3}). It

TABLE 6.2   Design Specifications of the SITO Inverter with Unified Resonant Frequency

Design Parameter	Value
Input voltage (V_{in})	5 V
Main switch frequency (f_{sw})	300 kHz
Resonant frequency (f_o)	100 kHz
Main inductor (L)	2.6 µH
ESR of the main inductor (measured at 300 kHz)	20 mΩ
Capacitor in the resonant tank (C_{o1}, C_{o2}, C_{o3})	0.37 µF
ESR of the resonant capacitor (measured at 100 kHz)	20 mΩ
Inductor in the resonant tank (L_{o1}, L_{o2}, L_{o3})	6.8 µH
ESR of the resonant inductor (measured at 100 kHz)	25 mΩ
Load resistor (R_{T1}, R_{T2}, R_{T3})	22 Ω

can be seen that S_{main} switches at 300 kHz. S_{out1}, S_{out2}, and S_{out3} switch at the same frequency of 100 kHz. A positive gate-to-source voltage (V_{GS}) is used to switch on the GaN transistor, whereas a negative V_{GS} is used to turn it OFF.

Figure 6.22 shows that the SITO inverter generates three nearly identical sinusoidal-like output voltages of the same peak amplitude and fundamental frequency. The measured RMS values of the first, second, and third output voltages are 6.04, 6.10, and 6.10 V, respectively. It is also observed that there is a phase difference of 120° between any two neighboring outputs. Figure 6.22 shows that the frequency of the main inductor current is three times that of the output voltage. Given an input power of 5.8 W, the total output power of the three outputs is measured to be 5.04 W. The

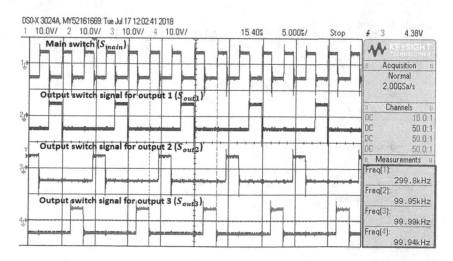

FIGURE 6.21   Measured waveforms of the gate drive signals of the main switch (S_{main}) and the three output switches (S_{out1}, S_{out2}, S_{out3}).

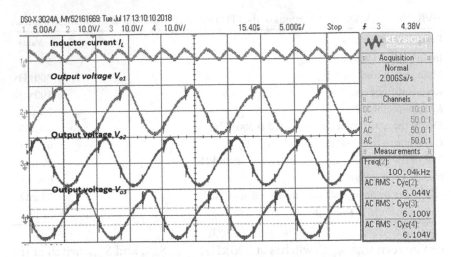

FIGURE 6.22 Measured waveforms of the main inductor current (I_L) and the three output voltages (V_{o1}, V_{o2}, V_{o3}) of the SITO inverter.

measured power efficiency of this GaN SITO inverter is around 86.9%, which is higher than that of its silicon counterpart.

In the second experiment, each output node is connected to a commercially available transmitting coil that is tightly coupled with a resistor-loaded receiving coil to mimic a real scenario of WPT. The output voltages are measured at the receiver side. It is clearly shown in Figure 6.23 that the measured voltages at the receiver output appear as smooth sine

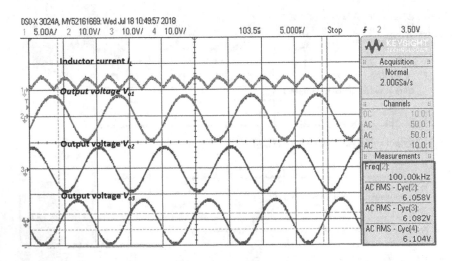

FIGURE 6.23 Measured waveforms of the main inductor current (I_L) and the three output voltages (V_{o1}, V_{o2}, V_{o3}) at the receiver output in a real WPT system.

waves with a fundamental frequency of 100 kHz. No noticeable distortions and voltage spikes are observed in the measured waveforms.

The total harmonic distortions (THDs) of the first, second, and third output voltages are measured to be 3.4%, 2.9%, and 3.0%, respectively. These THD values are sufficiently small. Hence, it is experimentally verified that the SITO inverter can simultaneously generate three clean sinusoidal-like output voltages of the same frequency at the receiver side in a practical WPT system.

In the third experiment, each output of the SITO inverter is loaded with a chassis mount resistor. The value of the load resistor for the first (or second) output is 22 Ω, whereas that for the third output is 47 Ω. Uniform on-time duty ratios are applied to the main switch (S_{main}). Figure 6.24 shows the measured waveforms of the main inductor current (I_L) along with the three output voltages (V_{o1}, V_{o2}, and V_{o3}). The measured RMS values of V_{o1} and V_{o2} are shown to be very close to each other (around 6 V) since both outputs have the same load resistance value. However, the measured RMS value of V_{o3} is around 7.33 V, which is higher than that of V_{o1} and V_{o2}. Notice that the main inductor current exhibits uniform peak values across switching cycles. In other words, the main inductor delivers the same amount of energy to each of the three output channels. Since the resonant inductor, resonant capacitor, and the resistive load are connected in parallel, a larger load resistance increases the current flowing into the resonant inductor and resonant capacitor. The fact that more energy is being transferred to the parallel LC tank causes the peak amplitude (and also the RMS value) of the oscillating voltage to increase.

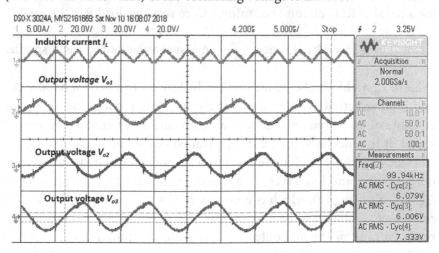

FIGURE 6.24 Measured waveforms of the main inductor current (I_L) and the three output voltages (V_{o1}, V_{o2}, V_{o3}) with nonuniform resistive loads.

So far, we have seen that the SITO inverter is capable of generating three sinusoidal-like output voltages with the same frequency of 100 kHz and reasonably low THD. The functionality of the SITO inverter using GaN transistors and isolated gate drivers is experimentally verified. In the following section, we will verify that the SITO inverter can also produce different output frequencies across the three outputs. Above all, the effectiveness of Type II, Type III, and modified Type III switching methods will be individually tested.

6.5.2 Different Output Frequencies

Regardless of the switching schemes, the three resonant tanks of the SITO inverter must have distinct values of resonant inductor and resonant capacitors so as to attain different output frequencies. Specifically, the values of the resonant inductor (L_{ox}) and resonant capacitor (C_{ox}) must be chosen in such a way that their product satisfies equation (6.2a–c), where $x = 1$, 2, and 3. The resonant inductor and resonant capacitor for the first, second, and third outputs are represented by (L_{o1}, C_{o1}), (L_{o2}, C_{o2}), and (L_{o3}, C_{o3}), respectively. Based on the commercially available discrete inductors and capacitors, there are many possible combinations of resonant inductor and resonant capacitor values that can produce the target resonant frequency. Yet in practice, it is preferable to keep a small circulating current in the LC resonant circuit at resonance in order to reduce the power loss due to parasitic resistances. By increasing the value of L_{ox} and reducing the value of C_{ox}, the effective reactance can be increased, and thus, the circulating circuit will be relatively small. On the other hand, for a parallel RLC circuit (assuming a pure resistive load R_L), the quality factor (Q_x) is given by $Q_x = R_{Lx}\sqrt{\dfrac{C_{ox}}{L_{ox}}}$. A larger value of C_{ox} and a smaller value of L_{ox} yield a higher Q factor, thereby leading to a narrow bandwidth and high selectivity of the resonant circuit. This reduces the harmonic contents of the output voltage so that it appears as a smoother sinusoidal signal. It is conceivable that a design tradeoff exists between high power efficiency and lower THD. We need to make a compromise between high output power and low THD. As a rule of thumb, the typical reactance of L_{ox} (i.e., $X_{Lx} = \omega_{ox}L_{ox}$), which is equal to that of C_{ox} (i.e., $X_{Cx} = 1/\omega_{ox}C_{ox}$) at resonance, is chosen to be a few ohms. For example, X_{L1} (or X_{C1}) for the resonant tank of the *first* output is chosen to be around 4.3 Ω. Given a particular resonant frequency, the appropriate values of resonant inductor and resonant capacitor can therefore be determined.

In principle, the value of the main inductor should be much larger than that of the resonant inductor. The theoretical analysis proves that when $L \gg L_o$, the frequency of the output voltage will be identical to the resonant frequency. In other words, the output voltage will theoretically be a pure sinusoidal signal with no distortion. Nonetheless, a larger value of the main inductor decreases the rising slope of the main inductor current (I_L) during the charging phase in DCM. This results in a smaller peak value of I_L for a fixed switching period. This will inevitably reduce the maximum amount of energy that can be stored in the main inductor in the charging phase (i.e., Mode 1). Thus, from the practical point of view, the main inductor value should be chosen in such a way that it is *smaller* than that of the resonant inductor in order to allow more energy to be transferred from the main inductor to each individual output. In this case, the main inductor is chosen to be about one third of the smallest resonant inductor. Table 6.3 contains the design specifications of the SITO boost inverter with different resonant frequencies. For the custom-made inductors, the core material is T35. As far as the main inductor ($L = 0.4\ \mu H$) is concerned, its coil has one turn and is made up of 240 strands of 38 AWG (0.1 mm) wires, where AWG stands for American Wire Gauge [15]. For the resonant inductor in the first output ($L_{o1} = 6.8\ \mu H$), its coil has a total of five turns, and it consists of 120 strands of 38 AWG (0.1 mm) wires. For the resonant inductor in the second output

TABLE 6.3 Design Specifications of the SITO Inverter with Different Resonant Frequencies

Design Parameter	Value
Input voltage (V_{in})	5 V
Main inductor (L)	0.4 μH
Load resistor (R_{T1}, R_{T2}, R_{T3})	22 Ω
First output	
Output frequency (f_{o1})	100 kHz
Capacitor in the resonant tank (C_{o1})	0.37 μF
Inductor in the resonant tank (L_{o1})	6.8 μH
Second output	
Output frequency (f_{o2})	200 kHz
Capacitor in the resonant tank (C_{o2})	0.23 μF
Inductor in the resonant tank (L_{o2})	2.78 μH
Third output	
Output frequency (f_{o3})	300 kHz
Capacitor in the resonant tank (C_{o3})	0.23 μF
Inductor in the resonant tank (L_{o3})	1.22 μH

(L_{o2} = 2.78 µH), its coil has a total of three turns, and it consists of 120 strands of 38 AWG (0.1 mm) wires. For the resonant inductor in the third output (L_{o3} = 1.22 µH), its coil has a total of two turns, and it consists of 120 strands of 38 AWG (0.1 mm) wires.

6.5.2.1 Verification of the Type II Switching Scheme

Since the Type II scheme is more relevant to a two-output inverter, only the first and second outputs of the SITO inverter are enabled. Simply put, the SITO inverter is reconfigured as a SIDO inverter. The target frequencies of the first and second output are 100 and 200 kHz, respectively. At first, each of the two output channels is loaded with a 22 Ω resistor. The main objective of this experiment is to verify the feasibility of this switching method. Figure 6.25 shows the measured waveforms of the gate drive signals of the main switch (S_{main}) as well as the first and second output switches (S_{out1} and S_{out2}).

Figure 6.26 shows the measured waveforms of the main inductor current and the two output voltages of the SIDO inverter. It clearly shows that the main inductor current (I_L) returns to zero toward the end of each switching cycle, thus indicating that the SIDO inverter operates correctly in DCM. The fundamental frequencies of the first and second outputs are measured to be around 100.14 and 200.08 kHz, which are in very close agreement with the corresponding target frequencies. The measured RMS

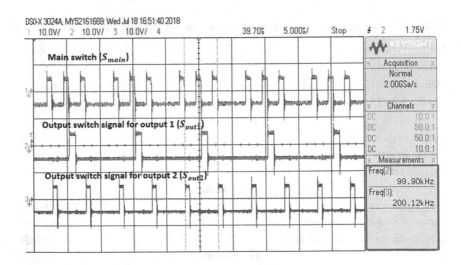

FIGURE 6.25 Measured waveforms of the gate drive signals of the main switch (S_{main}) and the two output switches (S_{out1}, S_{out2}) of the SIDO inverter with Type II switching scheme.

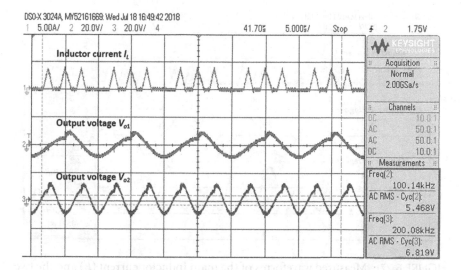

FIGURE 6.26 Measured waveforms of the main inductor current (I_L) and the two output voltages (V_{o1}, V_{o2}) of the SIDO inverter with Type II switching scheme.

values of the two output voltages are 5.468 and 6.819 V, respectively. Given an input power of 4.2 W, the combined output power of the two channels is measured to be 3.47 W, which yields a power efficiency of 82.6%.

Next, instead of loading the inverter with resistors, each of the two outputs is now connected to an off-the-shelf transmitting coil which is closely coupled with the receiving coil loaded with a resistor. This is to emulate real WPT in practical applications. The resonant circuit at the transmitting (Tx) coil and that at the receiving (Rx) coil are precisely tuned to be the same target frequency, which is also the fundamental component of the resonant frequency. The resonant Tx-Rx pair acts as a low-pass filter that helps to smooth out the output voltage at the receiver side. Figure 6.27 shows the measured waveforms of the main inductor current as well as the first and second output voltages measured at the receivers.

Figure 6.27 shows that the two output voltages measured at the receivers appear as clean and smooth sine waves with no noticeable distortions and voltage spikes. The measured frequencies of the two output voltages are 99.69 and 199.76 kHz, which are in close agreement with the corresponding target frequencies. The measured THD values for the first and second output voltages are 2.9% and 1.6%, respectively. Again, these THD values are sufficiently small. Hence, the experimental results confirm that the SIDO inverter with the Type II switching scheme is capable of producing two sinusoidal output voltages with different frequencies.

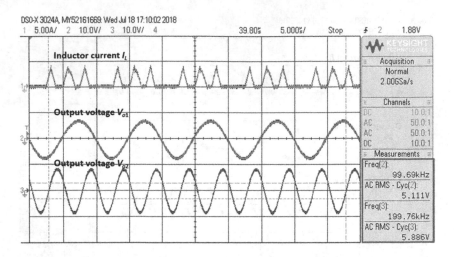

FIGURE 6.27 Measured waveforms of the main inductor current (I_L) and the two output voltages (V_{o1}, V_{o2}) at the receivers in practical WPT.

6.5.2.2 Verification of the Type III Switching Scheme

To verify the Type III switching method, all three outputs of the SITO inverter prototype are enabled by using the approach of multiple energizing phases. Only one output is enabled in every switching cycle. Figure 6.28 shows the measured waveforms of the gate drive signals of the main switch

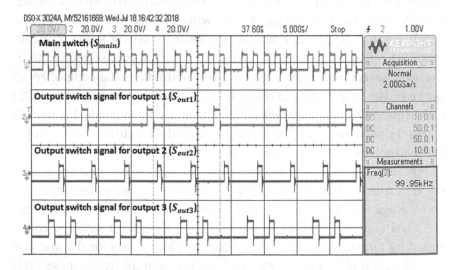

FIGURE 6.28 Measured waveforms of the gate drive signals of the main switch (S_{main}) and the three output switches (S_{out1}, S_{out2}, S_{out3}) of the SITO inverter with Type III switching scheme.

(S_{main}) and the three output switches (S_{out1}, S_{out2}, and S_{out3}). As can be seen clearly, all switches are opened in every sixth switching cycle because this serves as an idle cycle that is *not* being used by any of the outputs. Besides, S_{out1} is switched ON *once* every six switching cycles since the first output operates at 100 kHz. Likewise, S_{out2} (or S_{out3}) is ON *twice* every six switching cycles. Figure 6.29a and b shows the measured waveforms of the main inductor current (I_L) and the three output voltages (V_{o1}, V_{o2}, and V_{o3}). Even

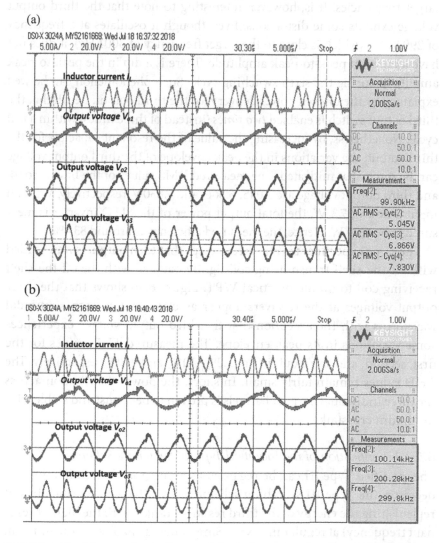

FIGURE 6.29 Measured waveforms of the inductor current I_L and the three output voltages (V_{o1}, V_{o2}, V_{o3}) of the SITO inverter, and the measured (a) RMS values and (b) frequencies of the three outputs are shown in the red box.

though both figures contain the same set of waveforms, Figure 6.29a is used to show the measured RMS values, whereas Figure 6.29b is used to show the measured frequencies of the three output voltages.

Figure 6.29 clearly shows that the first and second output voltages continue to appear as sinusoidal-like waveforms with constant peak-to-peak amplitude. The fundamental frequencies of the first and second output are 100.14 and 200.28 kHz, which agree very closely with the corresponding target frequencies. It is, however, interesting to note that the third output voltage exhibits some distortions. Even though it oscillates at a frequency of 299.8 kHz (which is close to the target frequency of 300 kHz), it does not have a uniform peak-to-peak amplitude. There is a dip in the peak-to-peak amplitude during the sixth switching cycle. Such phenomenon can be well explained by the uneven power delivery for the third output. Recall that the third output switch is enabled *two times* (instead of three times) within a full cycle. Nonetheless, by increasing the value of the resonant capacitor for the third output, the variations in the peak envelope of the third output voltage can be somewhat mitigated. The measured RMS values of the first, second, and third output voltage are 5.045, 6.866, and 7.830 V, respectively. With an input power of 7.3 W, the total output power of the SITO inverter is measured to be 6.09 W. Hence, the measured efficiency is around 83.4%.

The same experiment is repeated, except that each output is now loaded with off-the-shelf transmitting coil tightly coupled with the off-the-shelf receiving coil to mimic practical WPT. Figure 6.30 shows that the three output voltages at the receivers appear as clean and smooth sinusoidal waveforms with the exception that the third output voltage experiences some variations in its peak envelope. The measured THD values for the first, second, and third output are 3.2%, 2.1%, and 4.8%, respectively. The THD values remain fairly small. In short, the power distribution across the three individual output channels is proven to be robust and stable with minimum crosstalk.

6.5.2.3 Verification of the modified Type III Switching Scheme
The modified Type III can be regarded as an upgraded version of its predecessor, the Type III switching scheme. This enhanced scheme aims at replenishing the energy of the third resonant tanks (with the highest resonant frequency) at regular intervals. Simply put, it enables a more uniform power supply to the third output. Specifically, the third output is enabled *three* times in a full cycle. Figure 6.31 shows the switching sequence of all power switches. It clearly shows that the first output switch (S_{out1}) is enabled

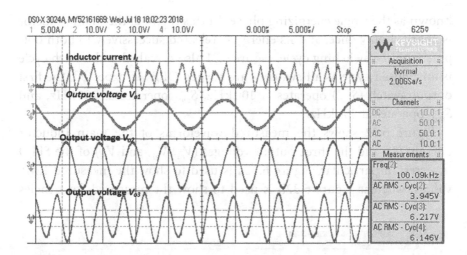

FIGURE 6.30 Measured waveforms of the main inductor current (I_L) and the three output voltages (V_{o1}, V_{o2}, V_{o3}) at the receivers in a practical WPT system.

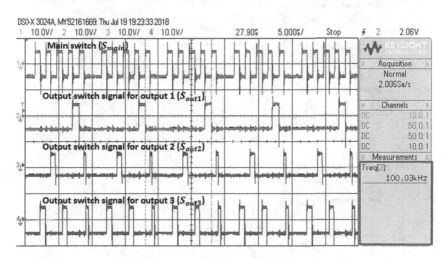

FIGURE 6.31 Switching sequence of the main switch (S_{main}) and the three output switches (S_{out1}, S_{out2}, S_{out3}) of the SITO inverter with modified Type III switching scheme.

once, the second output (S_{out2}) is enabled *twice*, and the third output (S_{out3}) is enabled *thrice* per full cycle. Most notably, at every fifth switching cycle, the third output switch (S_{out3}) is switched ON immediately after the second output switch (S_{out2}) is switched OFF. In other words, both the second and third output are sequentially enabled in the fifth switching cycle. This is indicative of a switching method of the conventional SIMO converter known as the single energizing phase. In this method, the main inductor is charged only once, and its energy is released successively to both outputs within a switching cycle. Figure 6.31 also reveals that the sixth cycle is an unused cycle as all switches are turned OFF. As expected, the first output switch (S_{out1}) operates at 100 kHz, S_{out2} operates at 200 kHz, and S_{out3} operates at 300 kHz.

Figure 6.32 shows the measured waveforms of the main inductor current (I_L) and the three output voltages (V_{o1}, V_{o2}, and V_{o3}) of the SITO inverter. In particular, Figure 6.32a is used to show the measured RMS values, whereas Figure 6.32b is used to show the measured frequencies

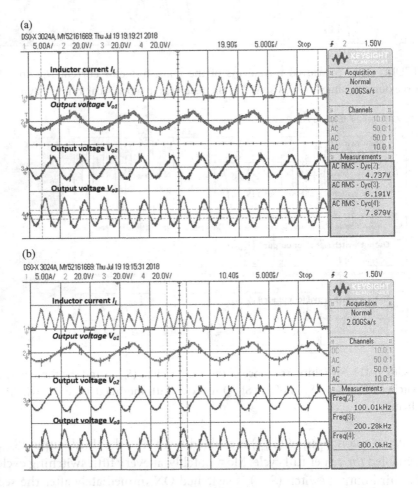

FIGURE 6.32 Measured waveforms of the inductor current I_L and the three output voltages (V_{o1}, V_{o2}, V_{o3}) of the SITO inverter, and the measured (a) RMS values and (b) frequencies of the three outputs are shown in the red box.

of the three output voltages. It is observed that the three output voltages remain sinusoidal-like. More importantly, the peak-to-peak amplitudes of the third output voltage are more uniform across cycles than that in Type III. In other words, the variations of the peak envelope of the third output voltage become smaller. The measured RMS values of the first, second, and third output voltages are 4.737, 6.191, and 7.879 V, respectively. The corresponding measured frequencies are 100.01, 200.28, and 300 kHz, which agree very closely with the corresponding target frequencies. Given an input power of 6.59 W, the total output power is measured to be 5.58 W, which yields a measured efficiency of 84.6%.

Instead of using pure resistive loads, the SITO inverter prototype is reconfigured in such a way that each output is connected to a pair of transmitting coil and a loaded receiving coil to imitate real WPT. Figure 6.33 shows the measured waveforms of the inductor current and the three output voltages at the receiver side. The experimental results confirm our expectation in that the three output voltages appear as smooth and stable sinusoidal waveforms with no cross-regulation. Moreover, the measured frequencies are in good agreement with the corresponding target values. Figure 6.33 also reveals that the peak-to-peak amplitude of the third output voltage (V_{o3}) is nearly constant across cycles, compared to that in Type III [see Figure 6.30]. This demonstrates the effectiveness of the modified Type III switching sequence. The measured THD values for the first, second, and third output voltages are 3.1%, 1.6%, and 2.7%, respectively. It is important

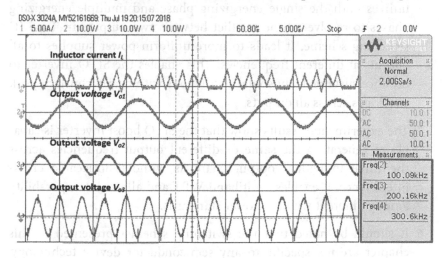

FIGURE 6.33 Measured waveforms of the main inductor current (I_L) and the three output voltages (V_{o1}, V_{o2}, V_{o3}) at the receivers in a practical WPT system.

to note that the reduction of THD of the third output is quite substantial (i.e., from 4.8% in Type III to 2.7% in modified Type III). In summary, it is experimentally verified that the modified Type III is superior to Type III in terms of signal integrity and robustness for multiband WPT.

6.6 SUMMARY OF KEY POINTS

1. A GaN-based SIMO inverter is introduced. Unlike its silicon counterparts, this inverter takes advantage of the absence of an intrinsic body diode in a GaN transistor as well as applying proper biasing between the gate-to-source voltage in order to eliminate reverse conduction across the GaN transistor.

2. An important attribute of the power stage of the GaN-based SIMO inverter is that it does *not* require any additional blocking diodes or back-to-back connected power transistors. It leads to a simplified circuit structure, reduced component count, smaller size, and higher efficiency, compared to its silicon counterparts.

3. It is analytically proven that by employing either Type III or modified Type III switching scheme, the SITO inverter can simultaneously produce three sinusoidal-like outputs of distinct frequencies.

4. Modified Type III is essentially a hybrid switching method, which utilizes both the single energizing phase and multiple energizing phases to resolve a time conflict between two outputs in Type III switching scheme. It leads to more uniform power supplies to all outputs at different frequencies. This enables the SITO inverter to produce sustained sinusoidal oscillations with nearly constant peak envelopes across all outputs.

5. The experimental results show that the SITO boost inverter is capable of generating the same or different output frequencies across the three individual output channels without noticeable cross-interference. It enables multiband WPT and allows interoperability among different wireless power standards.

6. It should be noted that the switching schemes presented in this chapter are *not* specific to any semiconductor device technology. Even though the Type II, Type III, and modified Type III switching schemes are experimentally verified using GaN-based SITO inverter,

it does *not* preclude other SITO inverters based on a different device technology from implementing any of these schemes. For example, silicon-based SITO inverters (using silicon power MOSFETs) can also be used to realize any of these switching methods.

REFERENCES

1. Y. Zhang, T. Lu, Z. Zhao, F. He, K. Chen, and L. Yuan, "Selective wireless power transfer to multiple loads using receivers of different resonant frequencies", *IEEE Trans. Power Electron.*, vol. 30, no. 11, pp. 6001–6005, Nov. 2015.

2. Y. J. Kim, D. Ha, W. J. Chappell, and P. P. Irazoqui, "Selective wireless power transfer for smart power distribution in a miniature-sized multiple-receiver system", *IEEE Trans. Ind. Electron.*, vol. 63, no. 3, pp. 1853–1862, Mar. 2016.

3. A. K. RamRakhyani and G. Lazzi, "On the design of efficient multi-coil telemetry system for biomedical implants", IEEE Trans. Biomed. Circuits Syst., vol. 7, no. 1, pp. 11–23, Feb. 2013.

4. W. Zhong and S. Y. R. Hui, "Auxiliary circuits for power flow control in multifrequency wireless power transfer with multiple receivers", *IEEE Trans. Power Electron.*, vol. 30, no. 10, pp. 5902–5910, Oct. 2015.

5. M. Q. Nguyen, Y. Chou, D. Plesa, S. Rao, and J. C. Chiao, "Multiple-inputs and multiple-outputs wireless power combining and delivering systems", *IEEE Trans. Power Electron.*, vol. 30, no. 11, pp. 6254–6263, Nov. 2015.

6. Y. J. Park, et al., "A triple-mode wireless power-receiving unit with 85.5% system efficiency for A4WP, WPC, and PMA applications", *IEEE Trans. Power Electron.*, vol. 33, no. 4, pp. 3141–3156, Apr. 2018.

7. F. Liu, Y. Yang, Z. Ding, X. Chen, and R. M. Kennel, "A multifrequency superposition methodology to achieve high efficiency and targeted power distribution for a multiload MCR WPT system", *IEEE Trans. Power Electron.*, vol. 33, no. 10, pp. 9005–9016, Oct. 2018.

8. D. Ahn and P. P. Mercier, "Wireless power transfer with concurrent 200-kHz and 6.78-MHz operation in a single-transmitter device", *IEEE Trans. Power Electron.*, vol. 31, no. 7, pp. 5018–5029, Jul. 2016.

9. C. Zhao and D. Costinett, "GaN-based dual-mode wireless power transfer using multifrequency programmed pulse width modulation", *IEEE Trans. Ind. Electron.*, vol. 64, no. 11, pp. 9165–9176, Nov. 2017.

10. Z. Pantic, K. Lee, and S. M. Lukic, "Multifrequency inductive power transfer", *IEEE Trans. Power Electron.*, vol. 29, no. 11, pp. 5995–6005, Nov. 2014.

11. J. Wu, C. Zhao, Z. Lin, J. Du, Y. Hu, and X. He, "Wireless power and data transfer via a common inductive link using frequency division multiplexing", *IEEE Trans. Ind. Electron.*, vol. 62, no. 12, pp. 7810–7820, Dec. 2015.

12. W. Jin, A. T. L. Lee, S.-C. Tan, and S. Y. R. Hui, "A gallium nitride (GaN)-based single-inductor multiple-output (SIMO) inverter with multi-frequency AC outputs", *IEEE Trans. Power Electron.*, vol. 34, no. 11, pp. 10856–10873, Nov. 2019.

13. Application Guide: "GN001 Application Guide: Design with GaN Enhancement mode HEMT", GaN Systems (2018). [Online] Available: https://gansystems.com/wp-content/uploads/2018/02/GN001_Design_with_GaN_EHEMT_180228.pdf.

14. The Qi WPT System Power Class 0 Specification, Part 4: Reference Designs, Wireless Power Consortium, Version 1.2.4, Jan. 2018.

15. Article: "American Wire Gauge", Wikipedia (2021). [Online] Available: https://en.wikipedia.org/wiki/American_wire_gauge.

PROBLEMS

6.1 Name the benefits of using the enhancement mode GaN transistors rather than silicon MOSFETs in SIMO inverters.

6.2 The output frequencies of a single-inductor three-output (SITO) inverter are 80, 160, 240 kHz, respectively. Determine the switching frequency of the main switch with the (a) Type II and (b) Type III switching sequence.

6.3 The design parameters of the SITO inverter in Problem 6.2 are given in Table 6.4. Suppose the SITO inverter operates in discontinuous conduction mode (DCM) with Type III switching sequence.

Table 6.4

Design Parameter	Value
Input voltage (V_{in})	5 V
Main inductor (L)	0.4 µH
First output	
Output frequency (f_{o1})	80 kHz
Capacitor in the resonant tank (C_{o1})	0.37 µF
Inductor in the resonant tank (L_{o1})	X µH
Load resistor (R_{L1})	22 Ω
Second output	
Output Frequency (f_{o2})	160 kHz
Capacitor in the resonant tank (C_{o2})	0.37 µF
Inductor in the resonant tank (L_{o2})	Y µH
Load resistor (R_{L2})	22 Ω
Third output	
Output frequency (f_{o3})	240 kHz
Capacitor in the resonant tank (C_{o3})	0.23 µF
Inductor in the resonant tank (L_{o3})	Z µH
Load resistor (R_{L3})	22 Ω

a. Find the theoretical values of the resonant inductors in the first, second and third output, respectively.

b. Create a circuit model of this SITO inverter. Use simulation software (e.g. Pspice, PSIM, Simulink, etc.) to plot the waveforms of the gate drive signals of the main switch and the output switches as well as the inductor current and the output voltages.

c. Obtain the THD of the first, second and third output voltages, respectively.

6.4

a. Suggest a method to reduce the distortion of the third output voltage obtained in Problem 6.3.

b. Perform simulation verification of the proposed method by plotting the waveforms of the gate drive signals of the main switch and the output switches as well as the inductor current and the output voltages.

c. Obtain the THD of the third output voltage.

6.5 For an ideal single-inductor three-output (SITO) boost inverter with identical output frequencies, the voltage at the first output (v_{o1}) is assumed to be a pure sinusoidal waveform with a phase angle of 0°, which can be represented as

$v_{o1}(t) = V_m \sin(\omega_o t)$, where V_m is the amplitude of the sinusoidal output voltage and ω_o is the angular resonant frequency.

a. Derive the relationship between V_m and the DC input voltage (V_{in}) of the SITO boost inverter operating in steady-state discontinuous conduction mode (DCM) in terms of the on-time duty ratio of the main switch for the first output (D_{11}) and that of the first output switch (D_{12}). State your assumptions.

b. Based on the result in part (a), express the relationship between V_m and the input voltage (V_{in}) of the SITO boost inverter operating in steady-state boundary condition mode (BCM) in terms of the on-time duty ratio of the main switch for the first output (D_{11}).

6.6

a. Propose a new switching sequence for a single-inductor three-output (SITO) boost inverter such that the first output and the second output are nearly in phase with each other (i.e. their phase difference is very small) while the third output is almost 180° out of phase with respect to the first output. Assume the three AC outputs have the same frequency.

b. Draw the ideal waveforms of the gate drive signals of the main switch and the three output switches, the inductor current, and the three output voltages of the SITO boost inverter.

c. Perform simulation verification of the proposed switching sequence.

6.7 A DCM SITO boost inverter produces three identical sinusoidal output voltages oscillating at the resonant frequency of f_o under balanced load condition. Each output of the SITO inverter is loaded with a resistor R_L of the same value.

a. Derive a general expression for the power efficiency of this SITO inverter in terms of the on-time duty ratio of the main switch (D_1), the on-time duty ratio of the output switch (D_2) and other relevant design parameters, where D_1 and D_2 are the same across all outputs in balanced load condition. State your assumptions.

b. Suppose the DCM SITO boost inverter in part (a) is designed with the following parameter values.

Input voltage $V_{in} = 5\,V$
Resonant frequency $f_o = 100\,kHz$
Main inductor $L = 2.6\,\mu H$
Resonant inductor $L_o = 6.8\,\mu H$
Resonant capacitor $C_o = 0.37\,\mu F$
Load resistance $R_L = 47\,\Omega$
On-time duty ratio of the main switch = 10%
On-time duty ratio of the output switch = 12%

Based on the result from part (a), determine the power efficiency of the SITO inverter.

CHAPTER 7

Single-Inductor Multiple-Output DC–AC/DC Hybrid Converters

7.1 INTRODUCTION

As we have seen in previous chapters, the concept of single-inductor multiple-output (SIMO) has evolved significantly in recent years. It first appeared in the literature as a low-power SIMO DC-DC converter. It can serve as embedded power supplies in wireless microsensors and portable electronic devices with distinct and diverse voltage requirements to achieve both high performance and extended battery life. SIMO DC-DC converter with high output power is made possible by operating it in pseudo-continuous conduction mode (PCCM), which allows a larger amount of energy stored in the main inductor to be delivered to each output successively. Owing to its compactness, cost-effectiveness, high efficiency, improved scalability, and flexibility, the SIMO topology is especially suited for driving multi-string LEDs. High-power SIMO-based LED drivers have emerged, which enable a single-stage AC-DC power conversion with reduced component count and smaller form factor. This is highly beneficial to LED lighting retrofits and other space-constrained lighting applications. To carry the concept of SIMO a step further, the feasibility of a SIMO DC-AC (inverter) topology is investigated. By combining the functions of a conventional SIMO DC-DC topology and the DC-AC resonant circuits in a proper

DOI: 10.1201/9781003239833-7

manner, a novel SIMO inverter topology is created. It enables a compact, low-cost, and efficient DC-AC power supply for real applications requiring a flexible number of independent AC loads. For instance, the SIMO inverter can act as a multidevice wireless charger in a multicoil wireless power transfer (WPT) system. Yet the SIMO topologies we have seen so far can perform only one specific type of power conversion, i.e., DC-DC, AC-DC, or DC-AC conversion. Naturally, a question comes to mind: Can SIMO be used to perform both DC-DC and DC-AC power conversions (the so-called hybrid conversion) at the same time? This chapter will provide clear answers to this question. But, to start with, we need to understand the motivation for building such a hybrid converter that can simultaneously supply AC and DC loads from a single DC power source.

Undoubtedly, WPT has entered into a new era that will fundamentally change the way we charge a myriad of electronic devices such as smartphones, tablets, wearables, sensors for Internet-of-Things or machine-to-machine services, medical implants, electric vehicles, etc. Existing wireless charging solutions are simply inadequate to cope with the diversity and increased battery capacity of future mobile gadgets. Indeed, both developers and manufacturers of WPT face unprecedented challenges to deal with a greater number of unidentical electronic devices with ever-growing power rating and the demand for fast and reliable charging. Recently, high-capacity hybrid chargers (or power banks) are made available that often include one Qi wireless charging pad and several USB charging ports (e.g., Type C, Lightning, micro-USB), allowing quick and convenient charging of multiple electronic devices at the same time. Apparently, a straightforward approach of realizing the hybrid chargers is to use separate power converters for DC-AC and DC-DC conversions. Figure 7.1a depicts a two-stage boost-derived hybrid converter (BDHC) [1,2]. It comprises a DC-DC boost converter connected in series with a voltage source inverter. This two-stage BDHC supplies two outputs, namely, one AC output and one DC output, from a single DC power source. A major drawback of this cascaded boost inverter topology is that it employs two separate power converters, which is simply too bulky, costly, and inefficient for practical applications. Consequently, a single-stage BDHC is reported in [3,4], as shown in Figure 7.1b. This topology is formulated by replacing the main switch of a traditional DC-DC boost converter with a single-phase (or three-phase) bridge network. Compared with its two-stage counterpart, the single-stage BDHC requires one fewer power device and uses only a single controller for power flow management. But still, it can supply only two loads (i.e., one AC and one DC). It also suffers from unstable standalone AC operation as it would produce large voltage spikes across the power switches.

To address the shortcomings of the BDHC topology, a modified BHDC is presented in [5–8] whose circuit topology is shown in Figure 7.1c. Notice that an additional antiparallel switch is connected across the diode to divert some of the storage energy from the capacitor to the AC load. Despite the

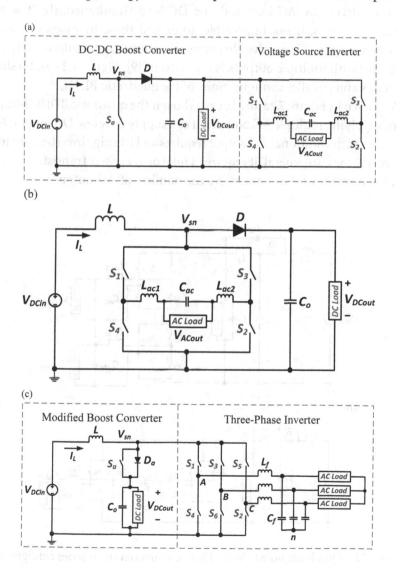

FIGURE 7.1   (a) Two-stage BDHC; (b) single-stage BDHC; (c) modified BDHC.

fact that it can achieve smooth and stable standalone AC operation, it suffers from a lower power efficiency than that of BDHC. As a result, a modified buck-boost-derived hybrid converter (MBBDHC) is reported in [8], which provides a wider range of buck or boost output voltage. By operating in a special mode called forced continuous current mode, this hybrid converter can achieve stable operation under standalone AC (or DC) load and hybrid loads. Nonetheless, these two variants of BDHC topologies can only drive one AC load and one DC load simultaneously. It would be rather tedious, if not impossible, to extend these topologies to more than two outputs. To resolve this issue, a quadratic boost-derived hybrid converter with multiple outputs is reported in [9]. Figure 7.2a and b show the series and parallel configurations of the quadratic BDHC.

As shown in Figure 7.2a and b, even though the quadratic BDHC is capable of supplying multiple AC outputs, it can supply only one DC output. Not only that, each additional AC output requires an H-bridge inverter with four power switches. In general, it requires a total of $4N$ power transistors, where N is the number of AC outputs. As the number of AC output increases, a

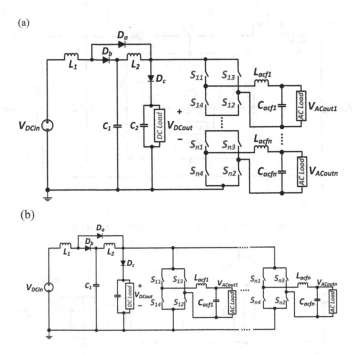

FIGURE 7.2 Quadratic boost-derived hybrid converter in (a) series configuration and (b) parallel configuration.

greater number of power transistors is required, which leads to increased component count, large form factor, high build-of-material cost, increased circuit and controller complexity, and lower power efficiency. In view of the limitations of the prior art, a SIMO DC-AC/DC hybrid converter with multiple AC and DC outputs is proposed in [10]. In the next section, the formulation of this hybrid converter topology will be thoroughly discussed.

7.2 CIRCUIT TOPOLOGY

The SIMO hybrid topology is created through proper integration of a generic DC-DC switching converter and a parallel network of LC resonant circuits and output filtering capacitors. The DC-DC converter transforms a DC voltage source into pulsating DC current. The DC current pulses flow from the main inductor to either the LC resonant circuit or the output filtering capacitor via the corresponding output switch in a time-multiplexed fashion. Fundamentally, the LC resonant circuit acts as the DC-AC stage that converts the pulsating DC current into sinusoidal-like output voltages. On the other hand, the output filtering capacitor is used to filter out the high-frequency ripples, thereby resulting in a nearly constant DC output voltage. Therefore, the SIMO hybrid topology can achieve a single-stage DC-to-AC/DC power conversion. By using a DC-DC buck converter as an example, Figure 7.3 provides a graphical illustration of the derivation of the SIMO buck hybrid topology. Alternatively, a boost (or buck-boost) DC-DC converter can also be used to derive the SIMO boost (or buck-boost) hybrid topology.

The ideal current source in Figure 7.3b for supplying each output channel can effectively be replaced by the main inductor (L), shown in Figure 7.3a, which acts like a current source. It generates pulsated DC current to each AC or DC output channel successively in a round-robin time-multiplexed manner. At the AC output channel, the LC resonant tank converts the pulsated DC current to a sinusoidal output voltage oscillating at the resonant frequency, which is given by

$$\omega_o = \frac{2\pi}{T_o} = \frac{1}{\sqrt{L_o C_o}} \tag{7.1}$$

where ω_o is the angular resonant frequency, T_o is the resonant period, L_o is the resonant inductor, and C_o is the resonant capacitor.

At the DC output channel, the output filtering capacitor is employed to mitigate the high-frequency ripples so as to yield a smooth DC output

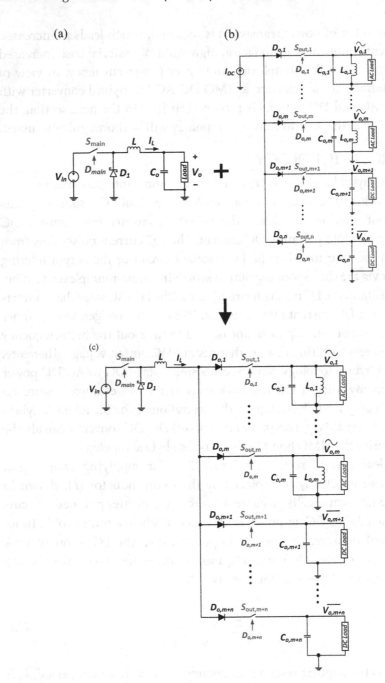

FIGURE 7.3 Derivation of a single-stage SIMO buck hybrid converter. (a) DC-DC buck converter; (b) ideal current source connected in series with a parallel network of LC resonant circuits; and (c) single-stage SIMO buck hybrid converter.

voltage. Just for the sake of illustration, Figure 7.3 assumes that silicon power MOSFETs are used in the SIMO hybrid converter. Hence, antiseries (blocking) diodes are required at the output branches to prevent any undesirable reverse conduction via the intrinsic body diode of the MOSFET. It should be pointed out that antiseries diodes are not only used at the AC output branches but also at the DC output branches as well in order to avoid potential cross-channel conduction between an AC and DC output or among the DC outputs. Apart from using an antiseries diode in series with a power MOSFET, it should be noted that a pair of back-to-back connected silicon power MOSFETs or even a GaN transistor can also be used.

Figure 7.3c shows a general SIMO-based buck hybrid converter with a total number of m AC outputs and n DC outputs. Compared to its prior art, a key advantage of this hybrid topology is that it supports any combination of AC and DC outputs. Only a single inductor L in the power stage is needed to provide steady energy supply to all outputs. Hence, a single-stage DC-to-AC/DC hybrid power conversion is realized. Without loss of generality, we will consider a single-inductor four-output (SIFO) buck-type hybrid converter with the quasi-hysteretic control scheme, as exemplified in Figure 7.4. In this SIFO buck hybrid converter, there are *two* AC outputs and *two* DC outputs, albeit other combinations of AC and DC outputs also being feasible. It transforms a single DC input voltage (V_{in}) into two sinusoidal-like AC output voltages ($V_{o1,ac}$, $V_{o2,ac}$) as well as two DC outputs ($V_{o3,dc}$, $V_{o4,dc}$). Instead of using antiseries diodes, back-to-back connected MOSFETs are used in this hybrid converter to achieve higher efficiency. It requires a total of *twelve* power MOSFETs, which includes a pair of main switches (Q_{m1}, Q_{m2}), four pairs of output switches [(Q_{o11}, Q_{o12}), (Q_{o21}, Q_{o22}), (Q_{o31}, Q_{out32}), (Q_{o41}, Q_{o42})], and a pair of freewheeling switches (Q_{fw1}, Q_{fw2}) across the main inductor. The current across the main inductor is annotated by I_L. Likewise, the output currents in the first, second, third, and fourth output branches are annotated as I_{o1}, I_{o2}, I_{o3}, and I_{o4}, respectively. These notations will be used consistently throughout this chapter.

In the power stage of the SIFO hybrid converter, each switch is realized as a pair of back-to-back-connected power MOSFETs. More specifically, the main power switch is implemented by a pair of back-to-back power MOSFETs (Q_{m1}, Q_{m2}). The two MOSFETs are connected in such a way that their source terminals are tied together so that their gate-to-source voltages are the same. This also allows the internal body diodes of the two MOSFETs to face away from each other. In addition, their gate terminals

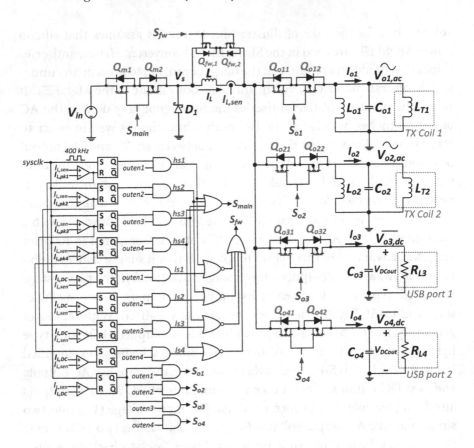

FIGURE 7.4   Circuit diagram of an SIFO buck hybrid converter.

are also tied together so that only a single gate driver is needed to drive the two MOSFETs. Imagine a situation when Q_{m2} is removed, leaving only Q_{m1} connected between the input voltage (V_{in}) and the switching node (V_s) in the SIFO converter. As soon as the instantaneous voltage of the first AC output voltage becomes greater than the input voltage (i.e., $V_{o1,ac} > V_{in}$), a reverse current will begin to flow from the first output channel to the input terminal via the body diode of Q_{m1} even when Q_{m1} is turned OFF and the first output is enabled. Therefore, Q_{m2} is needed to eliminate the unwanted reverse current. Likewise, the output switch at each output channel is realized by a pair of back-to-back MOSFETs to ensure a unidirectional current flow from the main inductor to each output as well as eliminating cross-regulation across the outputs. Imagine the case when the second back-to-back MOSFETs (i.e., Q_{o12}, Q_{o22}, Q_{o32}, Q_{o42}) of *all* output channels are removed. Assume the first output channel is enabled (i.e., Q_{o11} is switched ON). As soon as the instantaneous value of the first output voltage becomes less than the third output voltage (i.e., $V_{o1,ac} < V_{o3,dc}$), a cross-channel current will start to flow from the third output channel to the first output channel via the body diode of Q_{o31} even when Q_{o31} is switched OFF, thus resulting in an undesirable cross-regulation between the two outputs. A similar cross-channel conduction will also occur if the two DC outputs have *unequal* voltage values. Hence, the internal body diode of the second back-to-back MOSFET is required to block any unwanted conduction paths across the outputs under all circumstances.

It is rather interesting to note that the SIFO hybrid converter contains two back-to-back freewheeling MOSFETs (Q_{fw1}, Q_{fw2}) across the main inductor. This enables the inductor current to "recycle" between the main inductor and the freewheeling MOSFETs, thereby producing a positive DC offset value in the inductor current (I_L). Essentially, the SIFO converter operates in PCCM [10–15]. In PCCM, the hybrid converter can deliver a larger inductor current to each output than that in discontinuous conduction mode (DCM) while still achieving zero cross-regulation. The purpose of using two back-to-back freewheeling MOSFETs with their internal body diodes pointing towards each other is to prevent a short circuit between the switching node (V_s) and the output. Let us consider the following scenario. Suppose the first output is enabled by turning on its corresponding output MOSFETs (Q_{o11}, Q_{o12}) while the other outputs are disabled. In the absence of the second freewheeling MOSFET (Q_{fw2}), there exists a conduction path from V_s to $V_{o1,ac}$ via the internal body diode of Q_{fw1} when $V_s > V_{o1,ac}$ despite the fact that Q_{fw1} is switched OFF. In essence, the main inductor (L) is being short-circuited. Obviously, this is *not* the intended circuit behavior. As we can see, a pair of back-to-back connected freewheeling MOSFETs is necessary for this hybrid converter. Notice that the freewheeling MOSFETs could be removed if the converter were to operate solely in DCM at all times. In general, the SIMO-based buck hybrid converter requires a total of ($2N + 4$) power MOSFETs for PCCM operation or ($2N + 2$) power MOSFETs for DCM-only operation, where N is the total number of outputs.

A quasi-hysteretic feedforward control scheme is used to regulate the average inductor current of the SIFO converter in either PCCM or DCM. By controlling the peak values of the inductor current (I_L), the on-time duty ratios of main switch (S_{main}) corresponding to the four outputs can be individually adjusted, which in turn determine the peak (or RMS) values of the sinusoidal AC output voltages as well as the nominal values of the DC output voltages. Interested readers can refer to Chapter 5 (Section 5.3.3), which offers an in-depth discussion on the quasi-hysteretic control scheme.

7.3 OPERATING PRINCIPLE

In this section, the operating principle of a SIFO buck hybrid converter will be explained with the aid of the ideal waveforms of the key signals, the switching states along with the modes of operation. The same operating principle of the SIFO converter can be further extended to a single-inductor multiple-output (SIMO) buck hybrid converter with a total of N outputs.

It is assumed that the SIFO hybrid converter operates in PCCM at a fixed switching frequency and hence, a constant switching period (T_s). The first and second output voltages of this converter (V_{o1}, V_{o2}) are AC, whereas the third and fourth outputs (V_{o3}, V_{o4}) are DC. Since the converter has four outputs, a full resonant period (T_o) is *four* times the switching period (T_s), i.e., $T_o = 4T_s$. Figure 7.5a shows the ideal waveforms of the key signals such as the gate drive signals of the main switch (S_{main}), the output switches (S_{o1}, S_{o2}, S_{o3}, S_{o4}), and the freewheeling switch (S_{fw}) as well as the inductor current (I_L) and the four output voltages (V_{o1}, V_{o2}, V_{o3}, V_{o4}) spanning two full cycles. It shows the switching sequence of each output and the corresponding state transition. Each state is represented by (x, y), where x is the output number from 1 to 4, and y is the operating mode number from 1 to 3. For instance, state (2–1) denotes the second output and Mode 1. Figure 7.5a provides a graphical representation of the switching sequence of the outputs and also the state transitions (see the state annotations in the inductor current waveform). Figure 7.5b shows the equivalent circuit of each state. For ease of discussion, ideal switches and components are used in the equivalent circuit. The on-time duty ratios of S_{main} pertaining to the four individual outputs are denoted as D_{11}, D_{21}, D_{31}, and D_{41}, respectively. Since the converter operates in PCCM, I_L has a positive DC offset of $I_{L, DC}$. For a given value of $I_{L, DC}$, the on-time duty ratio of S_{main} is determined by the *peak limit* of the inductor current ($I_{L, pk}$). In fact, Figure 7.5a shows the general case of the SIFO converter with four distinct peak limits of the inductor current, which implies that the average inductor current delivered across the outputs will be different. Obviously, the higher the peak limit of the inductor current, the more energy can be delivered from

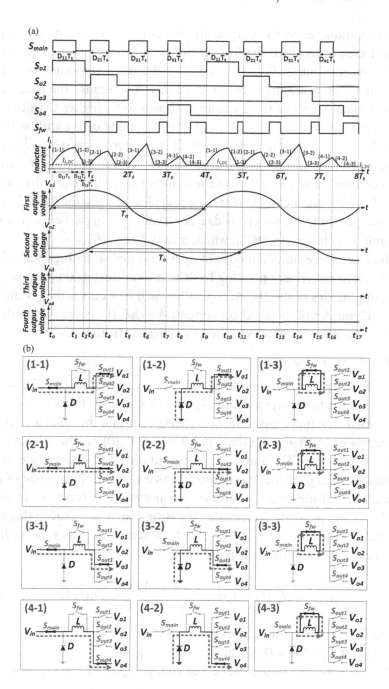

FIGURE 7.5 (a) Ideal waveforms of the gate drive signals of all switches, the inductor current, and the four output voltages; (b) equivalent circuits of the SIFO buck hybrid converter.

the main inductor to the output. In particular, since the peak value of the inductor current for the first output is shown to be *higher* than that for the second output (i.e., $I_{L,\,pk1} > I_{L,\,pk2}$), the peak amplitude (or RMS) value of V_{o1} is larger than that of V_{o2}. Likewise, because the peak value of the inductor for the third output is shown to be higher than that for the fourth output (i.e., $I_{L,\,pk3} > I_{L,\,pk4}$), the DC value of V_{o3} will be larger than that of V_{o4}. Consequently, the SIFO converter will produce different output power across its outputs. This is also referred to as the unbalanced load condition. The valley limit of the inductor current is equivalent to the DC offset of the inductor current ($I_{L,\,DC}$). Typically, the same valley limit is applied to all outputs, which yields a constant $I_{L,\,DC}$. In particular, the valley limit is set to zero in DCM, which means $I_{L,\,DC}$ is zero.

There are *three* unique modes of operation per switching cycle of the SIFO hybrid converter operating in PCCM. For illustration purpose, the operating modes of the DC-AC operation will be examined by using the first (AC) output of the SIFO converter. Likewise, the operating modes of DC-DC operation will be explained by using the third (DC) output.

7.3.1 DC-AC Operation

In Mode 1, e.g., from time t_0 to t_1 shown in Figure 7.5a, the main switch (S_{main}) and the first output switch (S_{o1}) are closed while the other switches are opened. The main inductor (L) is charged up, and the inductor current (I_L) ramps up with a rising slope $m_{11}(t) = [V_{in} - V_{o1}(t)]/L$, where $V_{o1}(t) = V_{m1} \sin(\omega_o t + \phi)$, where V_{m1} and ϕ are the amplitude and phase angle of the sinusoidal voltage of the first output. Notice that the rising slope of I_L is a function of time, and it resembles a concave down function. In other words, the rising slope *decreases* as the time t increases during this time interval. At the end of Mode 1 (e.g. $t = t_1$), I_L reaches a peak value of $I_{L,\,pk1}$, which can be mathematically represented as

$$I_{L,pk1} = m_{11}(t_1)D_{11}T_s + I_{L,DC} = \left[\frac{V_{in} - V_{o1}(t_1)}{L}\right]D_{11}T_s + I_{L,DC} \qquad (7.2)$$

where D_{11} is the on-time duty ratio of S_{main} for the first output, and $t_1 = nT_o + D_{11}T_s = (4n + D_{11})T_s$, and n is an integer.

At time $t = t_1$, the sinusoidal output voltage $V_{o1}(t_1)$ can be expressed as

$$V_{o1}(t_1) = V_{m1} \sin(\omega_o t_1 + \phi) = V_{m1} \sin\left(\frac{\pi D_{11}}{2} + \phi\right) \qquad (7.3)$$

The SITO converter makes a transition from Mode 1 to Mode 2 when $I_L = I_{L,pk1}$. Mode 1 for the first and second (AC) outputs are represented by (1–1) and (2–1), respectively, as depicted in Figure 7.5a and b.

In Mode 2, e.g., from time t_1 to t_2 shown in Figure 7.5a, the main switch (S_{main}) is switched OFF, whereas the first output switch (S_{o1}) remains ON. The other output switches remain OFF. The main inductor (L) discharges its energy to the first output. The energy stored in the main inductor is transferred to the first output. The inductor current (I_L) therefore ramps down with a slope of $m_{12}(t) = -\dfrac{V_{o1}(t)}{L}$ until it becomes equal to $I_{L,DC}$. But, since the time interval of Mode 2 is generally much shorter than a resonant period (i.e., $D_{12}T_s \ll T_o$), the variation in the sinusoidal $V_{o1}(t)$ is sufficiently small in this mode such that $m_{12}(t)$ can be approximated as follows.

$$m_{12}(t) \approx m_{12} = \frac{\overline{V_{o1,m2}}}{L} \tag{7.4}$$

where $\overline{V_{o1,m2}}$ is the average value of $V_{o1}(t)$ in Mode 2.

By using simple geometry, the average value of $V_{o1}(t)$ during Mode 2 can be obtained by integrating the total area under $V_{o1}(t)$ from $t_1 = D_{11}T_s$ to $t_2 = (D_{11} + D_{12})T_s$ and then dividing the definite integral by $D_{12}T_s$. This can be mathematically written as

$$\overline{V_{o1,m2}} = \frac{\displaystyle\int_{D_{11}T_s}^{(D_{11}+D_{12})T_s} V_{m1}\sin(\omega_o t + \phi)\,dt}{D_{12}T_s} \tag{7.5}$$

Since $T_s = \dfrac{T_o}{4} = \dfrac{\pi}{2\omega_o}$, equation (7.5) can be rewritten as

$$\overline{V_{o1,m2}} = \frac{2V_{m1}}{\pi D_{12}}(\cos\theta_1 - \cos\theta_2) \tag{7.6}$$

where $\theta_1 = \dfrac{\pi D_{11}}{2} + \phi$ and $\theta_2 = \dfrac{\pi(D_{11} + D_{12})}{2} + \phi$.

When I_L becomes equal to $I_{L,DC}$, the SIFO converter makes a transition from Mode 2 to Mode 3. Mode 2 for the first and second (AC) outputs are denoted by (1–2) and (2–2), respectively, as indicated in Figure 7.5a and b.

In Mode 3, e.g., from time t_2 to t_3 shown in Figure 7.5a, *all* switches, with the exception of the freewheeling switch (S_{fw}), are opened. During this time interval, the inductor current circulates between the main inductor and the freewheeling switch of the SIFO converter. It is maintained at a constant level ($I_{L,\,DC}$). Mode 3 of the first and second (AC) outputs are denoted by (1–3) and (2–3), respectively, as shown in Figure 7.5a and b. In particular, when the SIFO converter operates in DCM, all switches (including S_{fw}) will be opened. Hence, I_L will be reset to zero in Mode 3.

7.3.2 DC-DC Operation

In Mode 1, e.g., from time t_5 to t_6 shown in Figure 7.5a, the main switch (S_{main}) and the third output switch (S_{o3}) are closed, while the other switches are opened. The inductor current (I_L) ramps up linearly with a rising slope of $m_{31} = [V_{in}-V_{o3}]/L$, where V_{o3} is the third DC output voltage. Unlike DC-AC operation, the rising slope of I_L in DC-DC operation is constant since the output voltage is DC. At the end of Mode 1 (e.g., $t = t_6$), I_L eventually reaches its peak value $I_{L,pk3}$, which can be mathematically written as

$$I_{L,pk3} = m_{31}D_{31}T_s + I_{L,DC} = \left[\frac{V_{in}-V_{o3}}{L}\right]D_{31}T_s + I_{L,DC} \qquad (7.7)$$

where D_{31} is the on-time duty ratio of S_{main} for the third output.

In Mode 2, e.g., from time t_6 to t_7, the main switch (S_{main}) is opened, and the third output switch (S_{o3}) is closed, while the other switches remain opened. The main inductor releases its storage energy to the third output. Hence, I_L ramps down with a constant slope of $m_{32} = -\dfrac{V_{o3}}{L}$ until it is equal to $I_{L,\,DC}$. Finally, in Mode 3, e.g., from time t_7 to t_8, all switches other than the freewheeling switch are opened. This induces a freewheeling conduction path for the inductor current which has a constant DC value of $I_{L,\,DC}$.

It is vital to realize that the DC-AC and DC-DC operations of the SIFO converter are completely independent of each other. As a matter of fact, the operation of each output is fully decoupled in time due to the inherent nature of the time-multiplexing switching sequence in PCCM (or DCM). Each output can be separately enabled or disabled without affecting the other outputs. Specifically, standalone DC-AC operations can easily be achieved by enabling only the first and second (AC) outputs while disabling the third and fourth (DC) outputs. The SIFO converter is effectively transformed into a single-inductor dual-output (SIDO) DC-AC buck-type inverter. Figure 7.6 shows the circuit diagram of the SIDO buck-only inverter with *two* AC outputs. Indeed, this forms the basis for a new kind of SIMO inverter known as the SIMO buck inverter. Conversely, standalone DC-DC operations are also allowed by disabling the two AC outputs while enabling only the two DC outputs. The SIFO converter essentially becomes a SIDO DC-DC buck converter.

7.4 THEORETICAL CONSIDERATIONS OF SINUSOIDAL OSCILLATION

The theoretical underpinnings of sinusoidal oscillation at the outputs of SIMO boost-only inverter and SIMO buck-boost inverter are described in the preceding chapters. Yet the sinusoidal oscillation of a SIMO buck-only inverter remains to be fully understood. In this section, the sinusoidal oscillation at the AC output voltage of the SIMO buck hybrid converter in each mode of operation will be proven analytically. By enabling only the AC outputs, the functionality of the SIMO buck hybrid converter is virtually the same as that of a SIMO buck inverter. Hence, the proof of sinusoidal oscillation at the AC output of the SIMO buck hybrid converter is also applicable to that of the SIMO buck inverter. Without loss of generality, only the *first* output phase of the SIFO buck hybrid converter will be considered in the ensuing analysis.

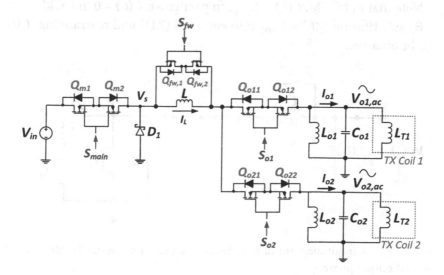

FIGURE 7.6 Circuit diagram of the SIDO buck-only inverter.

7.4.1 Mode 1—Proof of Sinusoidal Oscillation

Figure 7.7 shows the equivalent circuit of the SIFO buck inverter operating in Mode 1 for the first output phase.

Because S_{main} and S_{o1} are closed, the input voltage (V_{in}) is connected to the output resonant circuit via the main inductor (L). This provides a closed path for the inductor current (I_L). By applying KCL at the output node, I_L is the sum of the three output branch currents (i_{Co1}, i_{Lo1}, and i_{T1}), which can be expressed as

$$i_L(t) = i_{Co1}(t) + i_{Lo1}(t) + i_{T1}(t) \tag{7.8}$$

Equation (7.8) can be re-expressed in the frequency domain (s-domain) as

$$I_L(s) = I_{Co1}(s) + I_{Lo1}(s) + I_{T1}(s) \tag{7.9}$$

The voltage across the main inductor is given by

$$v_L(t) = V_{in} - v_{o1}(t) = L\frac{di_L(t)}{dt} \tag{7.10}$$

By taking the Laplace transform of equation (7.10), we have

$$V_L(s) = V_{in} - V_{o1}(s) = L[sI_L(s) - I_L(t_0)] \tag{7.11}$$

where $I_L(t_0)$ is the initial value of I_L in Mode 1 at $t = t_0$.

Note that in PCCM, $I_L(t_0) = I_{L,DC}$. In particular, $I_L(t_0) = 0$ in DCM.

By substituting $I_L(t_0) = I_{L,DC}$ into equation (7.11) and rearranging, $I_L(s)$ can be obtained as

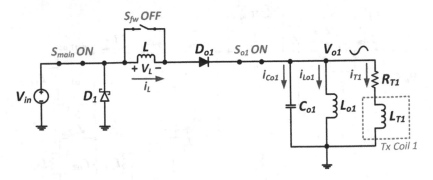

FIGURE 7.7 Circuit diagram of the SIFO buck hybrid converter in Mode 1 for the first output phase.

$$I_L(s) = \frac{V_{in} - V_{o1}(s) + LI_{L,DC}}{sL} \qquad (7.12)$$

The current across the resonant capacitor, denoted by $i_{Co1}(t)$, can be written as

$$i_{Co}(t) = C_{o1}\frac{dv_{Co1}(t)}{dt} \qquad (7.13)$$

where $v_{Co1}(t)$ is the voltage across the resonant capacitor (C_{o1}).

Observe that the resonant capacitor (C_{o1}) and the resonant inductor (L_{o1}) are connected in parallel. Thus, $v_{co1}(t) = v_{Lo1}(t) = v_{o1}(t)$. Hence, equation (7.13) can be reexpressed as

$$i_{Co}(t) = C_{o1}\frac{dv_{o1}(t)}{dt} \qquad (7.14)$$

Now, by applying Laplace transform on both sides of equation (7.14), we have

$$I_{Co}(s) = C_{o1}\left[sV_{o1}(s) - V_{o1}\right] \qquad (7.15)$$

where V_{o1} is the initial value of the output voltage, i.e., $V_{o1} = v_{o1}(t_0)$.

Likewise, the current across the resonant inductor (I_{Lo1}) can be expressed as

$$I_{Lo1}(s) = \frac{V_{o1}(s)}{sL_{o1}} + \frac{I_{Lo1}}{s} \qquad (7.16)$$

where I_{Lo1} is the initial value of the resonant inductor current, i.e., $I_{Lo1} = i_{Lo1}(t_0)$.

The current across the AC load, which consists of the resistor R_T in series with the inductance of the transmitting coil L_T, can be obtained as

$$I_{T1}(s) = \frac{V_{o1}(s) + L_{T1}I_{To1}}{R_{T1} + sL_{T1}} \qquad (7.17)$$

where I_{To1} is the initial value of the load current, i.e., $I_{To1} = i_{T1}(t_0)$.

Now, by substituting equations (7.12) and (7.15–7.17) into (7.9), we have

$$\frac{V_{in} - V_{o1}(s) + LI_{L,DC}}{sL} = C_{o1}\left[sV_{o1}(s) - V_{o1}\right] + \frac{V_{o1}(s)}{sL_{o1}} + \frac{I_{Lo1}}{s} + \frac{V_{o1}(s) + L_{T1}I_{To1}}{R_{T1} + sL_{T1}}$$

$$(7.18)$$

By assuming $L_{T1} \gg L_{o1}$ and rearranging the terms in equation (7.18), the following expression of $V_{o1}(s)$ can be obtained.

$$
V_{o1}(s) = V_{o1} \left(\frac{s}{s^2 + \dfrac{1}{LC_{o1}} + \dfrac{1}{L_{o1}C_{o1}}} \right) + \frac{V_{in}}{LC_{o1}} \left(\frac{1}{s^2 + \dfrac{1}{LC_{o1}} + \dfrac{1}{L_{o1}C_{o1}}} \right)
$$

$$
- \frac{I_{Lo1}}{C_{o1}} \left(\frac{1}{s^2 + \dfrac{1}{LC_{o1}} + \dfrac{1}{L_{o1}C_{o1}}} \right) + \frac{I_{L,DC}}{C_{o1}} \left(\frac{1}{s^2 + \dfrac{1}{LC_{o1}} + \dfrac{1}{L_{o1}C_{o1}}} \right)
$$

$$
\tag{7.19}
$$

$$
- \frac{sL_{T1}I_{To1}}{C_{o1}} \left(\frac{1}{s^2 + \dfrac{1}{LC_{o1}} + \dfrac{1}{L_{o1}C_{o1}}} \right) \left(\frac{1}{R_{T1} + sL_{T1}} \right)
$$

Let $\omega_1 = \sqrt{\dfrac{1}{LC_{o1}} + \dfrac{1}{L_{o1}C_{o1}}}$. Equation (7.19) can be reexpressed in terms of ω_1 in the following form.

$$
V_{o1}(s) = V_{o1} \left(\frac{s}{s^2 + \omega_1^2} \right) + \frac{V_{in}}{LC_{o1}} \left(\frac{1}{s^2 + \omega_1^2} \right) - \frac{I_{Lo1}}{C_{o1}} \left(\frac{1}{s^2 + \omega_1^2} \right) + \frac{I_{L,DC}}{C_{o1}} \left(\frac{1}{s^2 + \omega_1^2} \right)
$$

$$
- \frac{sL_{T1}I_{To1}}{C_{o1}} \left(\frac{1}{s^2 + \omega_1^2} \right) \left(\frac{1}{R_{T1} + sL_{T1}} \right)
\tag{7.20}
$$

Perform partial fraction expansion on $\left(\dfrac{1}{s^2 + \omega_1^2} \right)\left(\dfrac{1}{R_{T1} + sL_{T1}} \right)$, which is the last term on the R.H.S. of equation (7.20), and re-arranging, we have

$$
V_{o1}(s) = \left(\frac{C_{o1}V_{o1} + A_1}{C_{o1}} \right)\left(\frac{s}{s^2 + \omega_1^2} \right) + \left[\left(\frac{V_{in} - L(I_{Lo1} - I_{L,DC} - B_1)}{LC_{o1}} \right) \right]
$$

$$
\left(\frac{1}{s^2 + \omega_1^2} \right) + \frac{C_1}{L_{T1}} \left(\frac{1}{s + \dfrac{R_{T1}}{L_{T1}}} \right)
\tag{7.21}
$$

where $\qquad A_1 = -\dfrac{L_{T1}R_{T1}I_{To1}}{R_{T1}^2+\omega_1^2 L_{T1}^2}, \qquad B_1 = -\dfrac{L_{T1}^2 R_{T1}I_{To1}\omega_1^2}{R_{T1}\left(R_{T1}^2+\omega_1^2 L_{T1}^2\right)}, \qquad$ and

$$C_1 = \dfrac{L_{T1}^2 R_{T1}I_{To1}}{C_{o1}\left(R_{T1}^2+\omega_1^2 L_{T1}^2\right)}.$$

By taking inverse Laplace transform on both sides of equation (7.21), the output voltage in the time domain, $v_{o1}(t)$, can be obtained as follows.

$$v_{o1}(t) = a_1 \cos(\theta_1) + a_2 \sin(\theta_1) + a_3 e^{-\left(\frac{R_{T1}}{L_{T1}}\right)t} \qquad (7.22)$$

where $\quad a_1 = \dfrac{C_{o1}V_{o1}+A_1}{C_{o1}}, \quad a_2 = \dfrac{1}{\omega_1}\left[\dfrac{V_{in}-L\left(I_{Lo1}-I_{L,DC}-B_1\right)}{LC_{o1}}\right], \quad a_3 = \dfrac{C_1}{L_{T1}}, \quad$ and

$\theta_1 = \omega_1 t.$

In steady-state condition, the value of $e^{-\left(\frac{R_{T1}}{L_{T1}}\right)t}$ tends toward zero as the time t increases. This implies that the third term on the R.H.S. of equation (7.22) is negligible. Hence, equation (7.22) can be reexpressed as

$$v_{o1}(t) = \left(\sqrt{a_1^2+a_2^2}\right)\sin\left(\omega_1 t+\beta_1\right) \qquad (7.23)$$

where $\sin(\beta_1) = \dfrac{a_1}{\sqrt{a_1^2+a_2^2}}, \cos(\beta_1) = \dfrac{a_2}{\sqrt{a_1^2+a_2^2}},$ and $nT_s \le t < (n+D_{11})T_s.$

In theory, equation (7.23) shows that the output voltage in Mode 1 is a pure sinusoidal signal whose fundamental frequency is given by ω_1. In particular, when $L \gg L_{o1}$, $\omega_1 \approx \omega_o$. Simply put, when the value of the main inductor (L) is much larger than that of the output resonant inductor (L_{o1}), the frequency of the sinusoidal output voltage in Mode 1 is approximately the same as the resonant frequency of the resonant tank.

7.4.2 Mode 2—Proof of Sinusoidal Oscillation

Figure 7.8 depicts the equivalent circuit of the SIFO buck hybrid converter operating in Mode 2 for the first output phase.

In Mode 2, S_{main} is turned OFF, while S_{o1} remains ON. The inductor current (I_L) flows from ground via the freewheeling diode (D_1) and main inductor (L) to the output resonant tank. By invoking KCL at the output

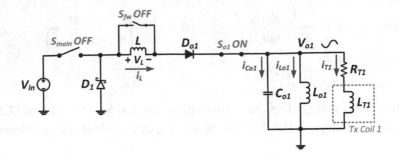

FIGURE 7.8 Circuit diagram of the SIFO buck hybrid converter in Mode 2 for the first output phase.

node, equations (7.8) and (7.9) continue to hold and thus, they will not be repeated here. The voltage across L in mode 2 is now given by

$$v_L(t) = 0 - v_{o1}(t) = L\frac{di_L(t)}{dt} \tag{7.24}$$

By invoking Laplace transform on equation (7.24), and re-arranging, we have

$$I_L(s) = \frac{LI_{L,pk1} - V_{o1}(s)}{sL} \tag{7.25}$$

where $I_{L,pk1}$ is the peak value of I_L for the first output.

By substituting equations (7.25) and (7.15–7.17) into (7.9), we have

$$\frac{LI_{L,pk1} - V_{o1}(s)}{sL} = C_{o1}\left[sV_{o1}(s) - V_{o1}\right] + \frac{V_{o1}(s)}{sL_{o1}} + \frac{I_{Lo1}}{s} + \frac{V_{o1}(s) + L_{T1}I_{To1}}{R_{T1} + sL_{T1}} \tag{7.26}$$

Assume $L_{T1} \gg L_{o1}$. $V_{o1}(s)$ can be obtained from equation (7.26) as follows.

$$V_{o1}(s) = V_{o1}\left(\frac{s}{s^2 + \dfrac{1}{LC_{o1}} + \dfrac{1}{L_{o1}C_{o1}}}\right) + \frac{I_{L,pk1}}{C_{o1}}\left(\frac{1}{s^2 + \dfrac{1}{LC_{o1}} + \dfrac{1}{L_{o1}C_{o1}}}\right)$$

$$-\frac{I_{Lo1}}{C_{o1}}\left(\frac{1}{s^2 + \dfrac{1}{LC_{o1}} + \dfrac{1}{L_{o1}C_{o1}}}\right) - \frac{sL_{T1}I_{To1}}{C_{o1}}\left(\frac{1}{s^2 + \dfrac{1}{LC_{o1}} + \dfrac{1}{L_{o1}C_{o1}}}\right)\left(\frac{1}{R_{T1} + sL_{T1}}\right) \tag{7.27}$$

Equation (7.27) can be further expressed in terms of ω_1 in the following form.

$$V_{o1}(s) = V_{o1}\left(\frac{s}{s^2+\omega_1^2}\right) + \frac{I_{L,pk1}}{C_{o1}}\left(\frac{1}{s^2+\omega_1^2}\right) - \frac{I_{Lo1}}{C_{o1}}\left(\frac{1}{s^2+\omega_1^2}\right)$$

$$-\frac{sL_{T1}I_{To1}}{C_{o1}}\left(\frac{1}{s^2+\omega_1^2}\right)\left(\frac{1}{R_{T1}+sL_{T1}}\right) \tag{7.28}$$

By performing partial fraction expansion on $\left(\dfrac{1}{s^2+\omega_1^2}\right)\left(\dfrac{1}{R_{T1}+sL_{T1}}\right)$,

which is the last term on the R.H.S. of equation (7.28), and simplifying, we have

$$V_{o1}(s) = \left(\frac{C_{o1}V_{o1} - A_2 L_{T1}I_{To1}}{C_{o1}}\right)\left(\frac{s}{s^2+\omega_1^2}\right) + \left(\frac{I_{L,pk1} - I_{Lo1} - B_2 L_{T1}I_{To1}}{C_{o1}}\right)$$

$$\left(\frac{1}{s^2+\omega_1^2}\right) - \frac{C_2 I_{To1}}{C_{o1}}\left(\frac{1}{s+\dfrac{R_{T1}}{L_{T1}}}\right) \tag{7.29}$$

where $A_2 = \dfrac{1}{R_{T1}+L_{T1}^2}$, $B_2 = \dfrac{L_{T1}}{R_{T1}+L_{T1}^2}$, and $C_2 = -\dfrac{I_{T1}}{R_{T1}+L_{T1}^2}$.

Now, by taking the inverse Laplace transform on both sides of equation (7.29), the output voltage in the time domain can be derived as

$$v_{o1}(t) = b_1 \cos(\theta_1) + b_2 \sin(\theta_1) + b_3 e^{-\left(\frac{R_{T1}}{L_{T1}}\right)t} \tag{7.30}$$

where $b_1 = \dfrac{C_{o1}V_{o1} - A_2 L_{T1}I_{To1}}{C_{o1}}$, $b_2 = \dfrac{1}{\omega_1}\left[\dfrac{I_{L,pk1} - I_{Lo1} - B_2 L_{T1}I_{To1}}{C_{o1}}\right]$, $b_3 = \dfrac{C_2 I_{To1}}{C_{o1}}$,

and $\theta_1 = \omega_1 t$.

The third term on the R.H.S. of equation (7.30) can be safely neglected. It is because as the value of t becomes much larger in steady-state condition, the value of $e^{-\left(\frac{R_{T1}}{L_{T1}}\right)t}$ tends to zero. As a result, equation (7.30) can be reduced to the following form.

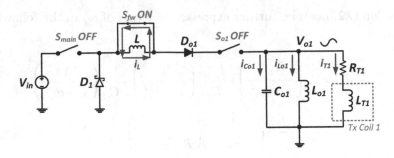

FIGURE 7.9 Circuit diagram of the SIFO buck hybrid converter in Mode 3 for the first output phase.

$$v_{o1}(t) = \left(\sqrt{b_1^2 + b_2^2}\right)\sin\left(\omega_1 t + \beta_2\right)$$ (7.31)

where $\sin(\beta_2) = \dfrac{b_1}{\sqrt{b_1^2 + b_2^2}}$, $\cos(\beta_2) = \dfrac{b_2}{\sqrt{b_1^2 + b_2^2}}$, and $(n+D_{11})T_s \le t <$

$(n+D_{11}+D_{12})T_s$

Equation (7.31) shows that the output voltage in Mode 2 is theoretically a pure sinusoidal signal with a fundamental frequency of ω_1. In the event that $L \gg L_{o1}$, $\omega_1 \approx \omega_o$. In other words, when the value of the main inductor is chosen to be much larger than that of the resonant inductor, the frequency of the sinusoidal output voltage will approximately be equal to the resonant frequency. It is interesting to note that the frequency of $v_{o1}(t)$ is the *same* between Mode 1 and Mode 2 because in either mode, the main inductor is connected to the output. This is a key attribute of the SIMO buck hybrid converter (or SIMO buck inverter), which distinguishes it from the SIMO boost and buck-boost counterparts.

7.4.3 Mode 3—Proof of Sinusoidal Oscillation

The equivalent circuit of the SIFO buck hybrid converter operating in Mode 3 for the first output phase is shown in Figure 7.9.

In this mode, both S_{main} and S_{o1} are in the open position, which implies that the output resonant tank is completely isolated from the main inductor L. The output resonant tank is basically in self-oscillation mode. S_{fw} is closed, which enables a circulating inductor circuit ($I_{L, DC}$) to flow across L. By applying KCL at the output node, we can write

$$i_{Co1}(t) + i_{Lo1}(t) + i_{T1}(t) = 0$$ (7.32)

Equation (7.32) can also be reexpressed in the s-domain, which is given by

$$I_{Co1}(s) + I_{Lo1}(s) + I_{T1}(s) = 0 \qquad (7.33)$$

By substituting equations (7.15)–(7.17) into (7.33), we have

$$C_{o1}\left[sV_{o1}(s) - V_{o1}\right] + \frac{V_{o1}(s)}{sL_{o1}} + \frac{I_{Lo1}}{s} + \frac{V_{o1}(s) + L_{T1}I_{To1}}{R_{T1} + sL_{T1}} \qquad (7.34)$$

Assume $L_{T1} \gg L_{o1}$. Based on equation (7.34), $V_{o1}(s)$ can be obtained as

$$V_{o1}(s) = V_{o1}\left(\frac{s}{s^2 + \dfrac{1}{L_{o1}C_{o1}}}\right) - \frac{I_{Lo1}}{C_{o1}}\left(\frac{1}{s^2 + \dfrac{1}{L_{o1}C_{o1}}}\right)$$

$$- \frac{sL_{T1}I_{To1}}{C_{o1}}\left(\frac{1}{s^2 + \dfrac{1}{L_{o1}C_{o1}}}\right)\left(\frac{1}{R_{T1} + sL_{T1}}\right) \qquad (7.35)$$

Hence, equation (7.35) can be reexpressed as

$$V_{o1}(s) = k_1\left(\frac{s}{s^2 + \omega_o{}^2}\right) - k_2\left(\frac{1}{s^2 + \omega_o{}^2}\right) + k_3\left(\frac{1}{s + \dfrac{R_{T1}}{L_{T1}}}\right) \qquad (7.36)$$

where $k_1 = V_{o1} - \dfrac{I_{To1}R_{T1}L_{o1}L_{T1}}{L_{T1}{}^2 + R_{T1}{}^2 L_{o1}C_{o1}}$, $k_2 = \dfrac{I_{Lo1}}{C_{o1}} + \dfrac{I_{To1}L_{T1}{}^2}{C_{o1}\left(L_{T1}{}^2 + L_{o1}C_{o1}R_{T1}{}^2\right)}$,

$k_3 = \dfrac{I_{To1}R_{T1}L_{o1}L_{T1}}{L_{T1}{}^2 + R_{T1}{}^2 L_{o1}C_{o1}}$, and $\omega_o = \dfrac{1}{\sqrt{L_{o1}C_{o1}}}$.

By taking the inverse Laplace transform on both sides of equation (7.36), the output voltage in the time domain can be expressed as

$$v_{o1}(t) = k_1\cos(\omega_o t) - \frac{k_2}{\omega_o}\sin(\omega_o t) + k_3 e^{-\left(\frac{R_{T1}}{L_{T1}}\right)t} \qquad (7.37)$$

As the value of t increases in steady-state condition, the value of $e^{-\left(\frac{R_{T1}}{L_{T1}}\right)t}$ tends toward zero, which suggests that the third term on the R.H.S. of equation (7.36) is negligible. Hence, equation (7.37) can be reduced to the following form.

$$v_{o1}(t) = c_1\cos(\theta_0) + c_2\sin(\theta_0) \qquad (7.38)$$

where $c_1 = k_1$, $c_2 = -\dfrac{k_2}{\omega_o}$, and $\theta_0 = \omega_0 t$.

Let $\sin(\beta_3) = \dfrac{c_1}{\sqrt{c_1{}^2 + c_2{}^2}}$, and $\cos(\beta_3) = \dfrac{c_2}{\sqrt{c_1{}^2 + c_2{}^2}}$.

Equation (7.38) can be rewritten as

$$v_{o1}(t) = \left(\sqrt{c_1{}^2 + c_2{}^2}\right)\sin\left(\omega_o t + \beta_3\right) \tag{7.39}$$

where $(n + D_{11} + D_{12})T_s \le t < (n+1)T_s$.

Equation (7.39) indicates that the output voltage in Mode 3 is a pure sinusoidal signal whose fundamental frequency is identical to the resonant frequency of the LC resonant circuit. In short, it is analytically proven that the SIMO buck hybrid converter, or SIMO buck inverter, can produce sinusoidal-like oscillation at the AC output. The angular frequencies of the output voltage are ω_1 in Mode 1 and Mode 2 and ω_o in Mode 3. This means that the AC output voltage exhibits some distortion when the converter makes a transition from Mode 2 and Mode 3 within the same cycle or from Mode 3 in the $(n-1)^{th}$ cycle to Mode 1 in the nth cycle.

7.5 SIMULATION VERIFICATION

The functionality of the SIFO buck hybrid converter with two AC and two DC outputs is verified by simulations using the PSIM software. Other SPICE simulators can also be used. Table 7.1 shows the design specifications of this hybrid converter.

TABLE 7.1   Design Specifications of the SIFO Buck Hybrid Converter

Design Parameter	Value
Input voltage (V_{in})	24 V
Switching frequency (f_{sw})	400 kHz
Output resonant frequency (f_o)	100 kHz
Main inductor (L)	3.4 µH
ESR of the main inductor (measured at 400 kHz)	58 mΩ
Capacitor in the resonant tank (C_{o1}, C_{o2})	0.56 µF
ESR of the resonant capacitor (measured at 100 kHz)	8 mΩ
Inductor in the resonant tank (L_{o1}, L_{o2})	4.5 µH
ESR of the resonant inductor (measured at 100 kHz)	9 mΩ
Filtering capacitor (C_{o3}, C_{o4})	33 µF
Load resistor (R_{T1}, R_{T2}, R_{T3}, R_{T4})	22 Ω

For standalone verification of the SIFO converter, each individual output is connected to a pure resistive load of 22 Ω. To start with, the SIFO converter operating in PCCM and balanced load condition is investigated. Under this operating condition, the two AC output voltages (V_{o1}, V_{o2}) should have the same frequency and RMS (or peak) values. Besides, the nominal values of the DC output voltages (V_{o3}, V_{o4}) should be the same, which are also equal to the RMS value of the AC output. Figure 7.10 shows the corresponding simulation results. It includes the simulated waveforms of the key signals of the SIFO hybrid converter such as the inductor current, the four output voltages as well as the gate drive signals of the main switch, the output switches, and the freewheeling switch.

The simulation results confirm that the SIFO hybrid converter is capable of producing two sinusoidal-like AC output voltages with a fundamental frequency of 100 kHz. The RMS values of the first and second (AC) output voltages are 7.91 and 7.93 V, respectively. The nominal values of the third and fourth (DC) output voltages are 7.89 and 7.91 V, respectively. Also, Figure 7.10 reveals that there is some distortion near the peak of the sinusoidal-like AC output voltage. Such sinusoidal distortion is highlighted in Figure 7.11. This is largely attributed to the fact that the value of the main inductor ($L = 3.4$ μH) is in the same order of magnitude as that of the resonant inductor ($L_o = 4.5$ μH). As predicted by the theoretical model

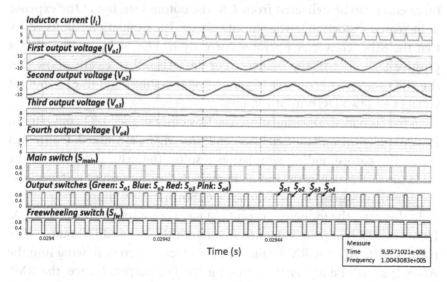

FIGURE 7.10 Simulation results of SIFO hybrid converter in PCCM and balanced load condition.

in Section 7.4, the frequency in Mode 1 (or Mode 2) will be higher than that in Mode 3 (i.e., $\omega_1 \approx 1.528\ \omega_o$). So, the distortion of the AC output voltage is caused by a change in frequency during mode transition.

Yet by making $L \gg L_o$, the frequency in Mode 1 (or Mode 2) will become nearly the same as that in Mode 3, i.e., $\omega_1 \approx \omega_o$. Hence, the distortion in the AC output voltage can be minimized. As an illustration, suppose the value of L is increased more than sevenfold (i.e., from 3.4 to 25 µH) while that of L_o remains unchanged. This implies $\omega_1 \approx 1.089\omega_o$, which means that the frequency in Mode 1 (or Mode 2) is fairly close to the frequency in Mode 3. Figure 7.12 shows the corresponding simulation results. The two sinusoidal-like output voltages appear much smoother and cleaner to the extent that the distortion is almost unnoticeable as compared with the corresponding waveforms in Figure 7.11. It is therefore envisaged that by using a larger value of L, the frequency of the output voltage in Mode 1 or Mode 2 will move closer to the resonant frequency, thereby mitigating the sinusoidal distortion in the AC outputs.

Nonetheless, the use of a larger main inductor value will significantly restrict the maximum amount of energy that can be stored in the main inductor during Mode 1 as the inductor current will rise at a slower rate. To achieve a higher output power, the value of L should *not* be too large. In fact, the chosen value of L is actually smaller than that of L_o, which allows more energy to be delivered from L to the output load but at the expense of increased distortion in the output voltage. The simulation results verify that the SIFO converter operates correctly in PCCM by maintaining a positive DC offset in the inductor current. Also, they clearly show that the SIFO converter is capable of generating two sinusoidal-like AC output voltages and two DC output voltages.

In the second set of simulation, the feasibility of the SIFO converter in PCCM under the *unbalanced* load condition will be examined. Figure 7.13 shows the corresponding simulation results. The critical signals of the SIFO converter such as the inductor current, the four output voltages, and the gate-drive signals of all switches are shown in this figure.

In this case, the inductor current (I_L) has a constant DC offset of 4.5 A. However, *different* peak values of I_L between the AC and DC outputs are used. Specifically, the RMS value of the inductor current flowing into the AC output will be *higher* than that for the DC output. Hence, the RMS value of the AC output voltage is expected to be *larger* than the nominal value of the DC output voltage. It turns out that the simulated RMS values

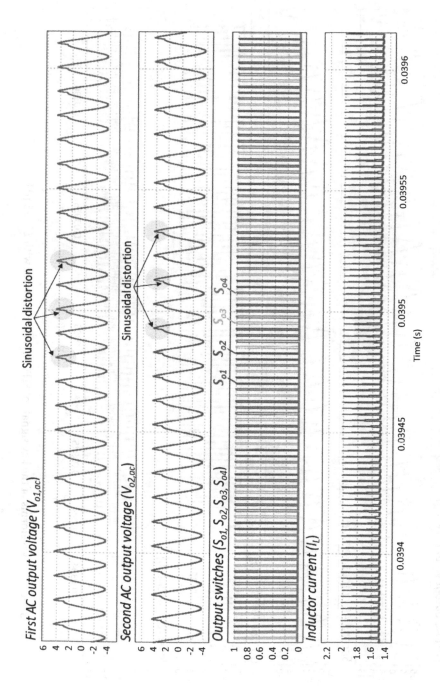

FIGURE 7.11 Simulated waveforms of the key signals of SIFO hybrid converter with $L = 3.4\ \mu H$.

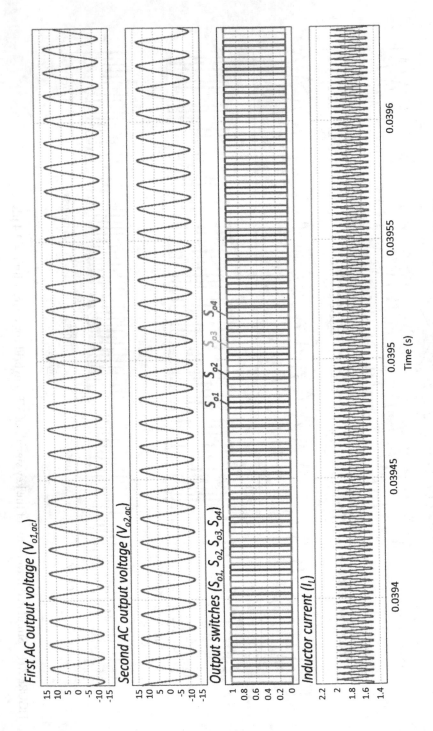

FIGURE 7.12   Simulated waveforms of the key signals of SIFO hybrid converter with $L = 25\ \mu H$.

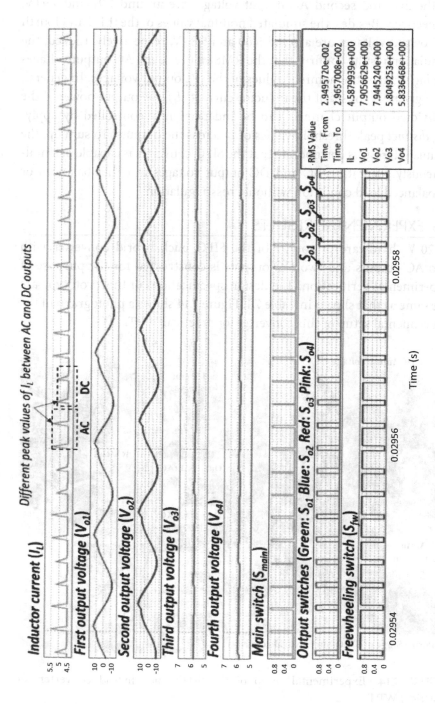

FIGURE 7.13   Simulation results of SIFO hybrid converter in PCCM and unbalanced load condition.

of the first and second AC output voltages are around 7.91 and 7.94 V, respectively. Besides, the simulated nominal values of the third and fourth DC output voltages are around 5.80 and 5.83 V, respectively. Indeed, the simulation results confirm that the RMS values of the AC output voltages are *higher* than the nominal values of the DC output voltages. In general, for a given DC offset of the inductor current (I_L), the output power of the individual output channels can be independently controlled by applying *distinct* peak threshold values of I_L across the outputs. To sum up, the simulation results corroborate that the SIFO converter is capable of simultaneously generating AC and DC output voltages in either balanced or unbalanced load condition without cross-regulation.

7.6 EXPERIMENTAL RESULTS

A 20-W hardware prototype of the SIFO buck hybrid converter with two AC outputs and two DC outputs is constructed for the purpose of experimental verification. The design specifications of this prototype are the same as that shown in Table 7.1. Figure 7.14 shows a photograph of the experimental setup of this converter for practical WPT.

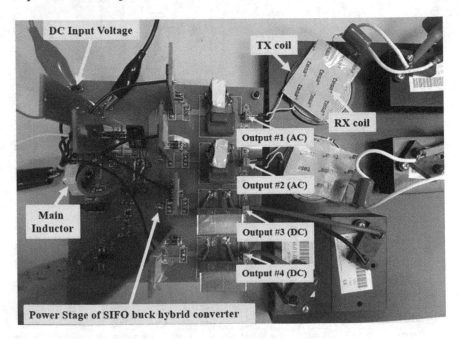

FIGURE 7.14 Experimental setup of the SIFO buck hybrid converter for practical WPT.

The power stage of the SIFO converter is implemented on a two-layer main printed circuit board (PCB) with a copper thickness of 2 oz for minimizing the trace resistance. The main PCB consists of active components (e.g., power MOSFETs, current-sense operational amplifiers, etc.) and passive components (e.g., main inductor, resonant inductors/capacitors, filtering capacitors, etc.). Six isolated gate drivers are used. Each isolated gate driver is implemented on a small auxiliary PCB, which is mounted vertically to the main board. The part numbers of the key components used in the prototype are listed in Table 7.2.

The main inductor and the resonant inductors are custom-made in order to minimize the equivalent series resistance for achieving higher efficiency. The core material of the main inductor is N41. Its coil has a total of 4 turns (with air gap) and is made up of 240 strands of AWG 38 (0.1 mm) Litz wire. On the other hand, the core material of the resonant inductor is T35. Its coil has a total of 8 turns and consists of 60 strands of AWG 38 (0.1 mm) Litz wire. The values of the resonant inductor and resonant capacitor are chosen in such a way that equation (7.1) is satisfied. Obviously, there are many possible combinations of resonant inductor (L_o) and resonant capacitor (C_o) values that can produce the same resonant frequency. Generally speaking, it is desirable to minimize the circulating current in the LC resonant tank at the resonance so as to reduce the conduction loss due to the parasitic resistances. This can be accomplished by means of a *larger* value of L_o and a *smaller* value of C_o so as to increase the overall reactance

TABLE 7.2   List of Key Components Used in Prototype

Component	Part Number
Power MOSFET	IPP083N10N5AKSA1
Gate driver	SI8261BAC-C-IS
Current-sensing resistor	MP930-0.10-1%
Main inductor (L)	Custom-made
Resonant inductor (L_{o1}, L_{o2})	Custom-made
Resonant capacitor (C_{o1}, C_{o2})	ECW-F4564HL
Filtering capacitor (C_{o3}, C_{o4})	B32564J1336K000
Transmitting (TX) coil	760308100110
Receiving (RX) coil	760308103211
Output resistor ($R_{L1}, R_{L2}, R_{L3}, R_{L4}$)	RCH25S22R00JS06
Operational amplifier	OPA354
Comparator	AD8611
Xilinx FPGA	Spartan-3E (XC3S250E)

value, thereby leading to a smaller circulating current. On the other hand, the quality factor of a parallel RLC circuit is given by $Q = R_L \sqrt{\dfrac{C_o}{L_o}}$, where R_L is the resistive load. The use of a larger value of C_o and a smaller value of L_o yields a higher Q factor, which results in a narrower bandwidth and higher selectivity of the resonant circuit. This is favorable because a higher Q factor helps reducing the harmonic contents of the AC output voltage, hence resulting in a cleaner and smoother sinusoidal waveform with very low distortion. It becomes clear that there exists a design tradeoff between higher power efficiency and lower total harmonic distortion (THD). So, as a general rule of thumb, the typical reactance of L_o (i.e., $X_L = \omega_o L_o$), which is the same as that of C_o (i.e., $X_C = 1/\omega_o C_o$) at resonance, is chosen to be a few ohms. For example, X_L (or X_C) at the resonance frequency (ω_o) is chosen to be around 2.8 Ω in this case. Hence, for a given ω_o, the proper values of L_o and C_o can be determined.

In principle, the value of the main inductor should be much greater than that of the resonant inductor (i.e., $L \gg L_o$) so that the frequency of the output voltage becomes identical to the resonant frequency under all modes of operation. Essentially, the AC output voltage of the SIMO hybrid converter will ideally be a pure sinusoidal signal. Nevertheless, a larger value of L reduces the rising slope of the main inductor current (I_L) during the charging phase (Mode 1), which unavoidably leads to a *smaller* peak value of I_L for a given switching period. This will reduce the maximum amount of energy stored in L that can be delivered to each output. Hence, from the practical perspective, the value of L is often chosen to be comparable to (or slightly smaller than) L_o in order to allow more energy to be delivered to each output while still maintaining a sufficiently low THD. In the hardware prototype of the SIFO buck hybrid converter, $L \approx (3/4)L_o$.

In reality, this SIFO converter can act as a multidevice hybrid charger (or power bank) comprising two wireless charging pads and two wired (USB) charging ports. Rather than connecting to a resistive load, each AC output is loaded with a pair of off-the-shelf transmitting coil and the corresponding off-the-shelf receiving coil to form a real inductive link for practical WPT. The single-layer transmitting coil has a total inductance value of 24 μH, a DC resistance (DCR) of 70 mΩ, and a rated current value of 6 A. The single-layer receiving coil has a total inductance value of 7.3 μH, a DCR of 200 mΩ, and a rated current of 2.5 A. On the other hand, each of the two DC outputs of the SIFO converter is connected to a 22-Ω resistor. A very small 50 mΩ current sense resistor is connected in series with the main inductor. The voltage across this sensing resistor, which is proportional to the current across the main inductor, is amplified by a high-speed operational amplifier with a unity-gain bandwidth of 200 MHz. Subsequently, a 4-ns fast comparator is used to compare the output voltage of the op amp against the peak (or valley) threshold for each output channel and then generate the peak-crossing and valley-crossing detection signals which are fed into the digital controller. The digital controller is implemented using Xilinx Spartan-3E FPGA because of its relatively high clock speed. It is mainly used to realize the quasi-hysteretic control scheme for regulating the inductor current. Based on the peak-crossing and valley-crossing input signals together with the internal clock signals, the digital controller determines the switching sequence by generating the proper logic signals for switching ON/OFF the power MOSFETs via the gate drivers.

7.6.1 Typical Operation (Two AC and Two DC Outputs)

In the first experiment, the hardware prototype of SIFO converter is configured in such a way that it can produce two AC output voltages as well as two DC output voltages simultaneously. Basically, *all* outputs are enabled. As a starting point, each output is loaded with a 22-ohm resistor. A balanced load condition is experimentally verified. Since the oscilloscope being used has only four channels, a maximum of four waveforms can be captured at any time instance. Figure 7.15a shows the measured waveforms of the inductor current (I_L), the first output voltage (V_{o1}), the second output voltage (V_{o2}), and the third output voltages (V_{o3}). In Figure 7.15b, the first three waveforms are the same as Figure 7.15a, except that the last one is the fourth output voltage (V_{o4}).

7.6.2 Standalone AC Operation

In the second experiment, the proposed converter operates in standalone AC mode. This means only the two AC outputs of the SIFO converter are enabled. The SIFO converter essentially behaves like an SIDO inverter. Figure 7.16a shows the measured waveforms of the inductor current (I_L) and the gate drive signals of the main switch (S_{main}), the first output switch, and the second output switch (S_{o1}, S_{o2}). For the sake of completeness, the gate-drive signals of the third and fourth output switches (S_{o3}, S_{o4}) are shown in Figure 7.16b, while that of the freewheeling switch (S_{fw}) is also shown in Figure 7.16c. The key to achieving this standalone AC mode is to enable S_{main} only for the first and second (AC) outputs. Hence, the effective switching frequency of S_{main} is 200 kHz (instead of 400 kHz). Also, the two

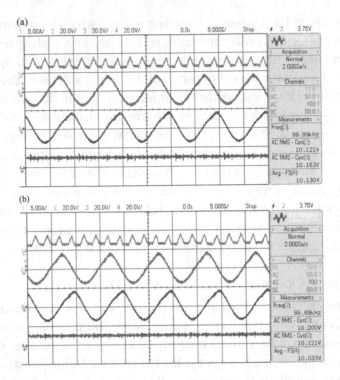

FIGURE 7.15 Measured waveforms of the inductor current (I_L), the first and second output voltages (V_{o1}, V_{o2}), and (a) the third output voltage (V_{o3}); (b) the fourth output voltage (V_{o4}).

DC outputs are completely disabled by turning OFF S_{o3} and S_{o4} at all times [see Figure 7.16b].

Figure 7.17a shows the measured waveforms of I_L, V_{o1}, V_{o2}, and V_{o3}, and Figure 7.17b shows the measured steady-state waveform of V_{o4}. It is experimentally tested that the SIFO hybrid converter produces sinusoidal-like AC voltages at the first and second outputs while producing *zero* DC voltages at the third and fourth outputs. The fundamental frequency of the first output is measured to be around 100 kHz. It clearly demonstrates that this hybrid converter is capable of operating in standalone AC mode due to the independent operation of the AC and DC outputs. The total output power of the two AC outputs is measured to be around 10.5 W.

7.6.3 Standalone DC Operation

After validating the standalone AC operation, it becomes natural for us to investigate the standalone DC operation. In this experiment, only the two DC outputs are enabled. Basically, the SIFO converter acts like a SIDO

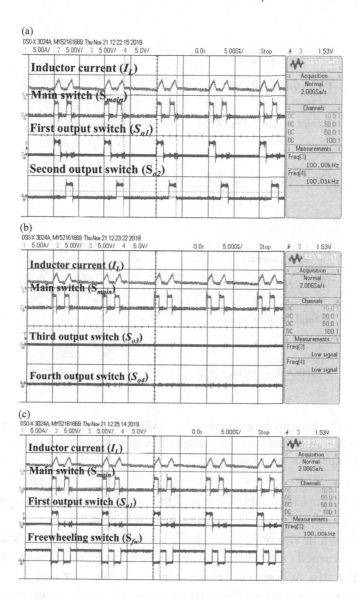

FIGURE 7.16 Measured waveforms of the inductor current (I_L), the main switch (S_{main}), and (a) the first and second output switches (S_{o1}, S_{o2}); (b) the third and fourth output switches (S_{o3}, S_{o4}); (c) the first output switch and the freewheeling switch (S_{o1}, S_{fw}).

DC-DC converter. Figure 7.18a contains the measured waveforms of I_L and the gate-drive signals of S_{main}, S_{o3}, and S_{o4}. It shows that S_{main} is switched ON only for the third and fourth outputs. Figure 7.18b shows that the first and second output switches (S_{o1}, S_{o2}) remain OFF all the time because the

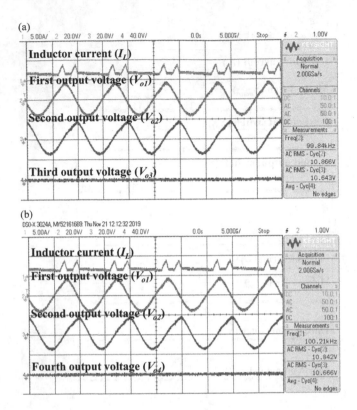

FIGURE 7.17 Measured waveforms of the inductor current (I_L), the first output voltage (V_{o1}), the second output voltage (V_{o2}), and (a) the third output voltage (V_{o3}); (b) the fourth output voltage (V_{o4}).

two AC outputs are disabled. Figure 7.18c shows the switching sequence of the freewheeling switch (S_{fw}) with respect to S_{main} and S_{o3}.

As shown in Figure 7.18a and c, S_{fw} is turned ON immediately after S_{o3} (or S_{o4}) is turned OFF. It is important to note that S_{fw} remains closed (ON) when the two AC outputs are completely disabled for two consecutive switching cycles. During this extended idle phase, the inductor current (I_L) continues to "freewheel" through S_{fw} and the main inductor. Since this freewheeling cycle is relatively long, the conduction loss contributed by the turn-on resistance of S_{fw} and the DCR of the main inductor becomes more pronounced, thus causing a gradual decrease in I_L.

Figure 7.19a shows the measured waveforms of the inductor current (I_L), the third output voltage (V_{o3}), the fourth output voltage (V_{o4}), and the first output voltage (V_{o1}). Figure 7.19b looks similar to Figure 7.19a, except that

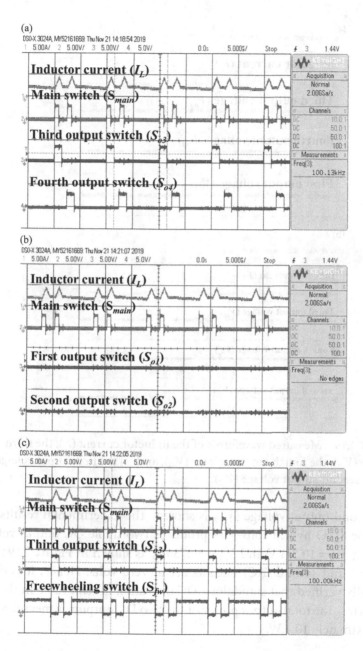

FIGURE 7.18 Measured waveforms of the inductor current (I_L), the main switch (S_{main}), and (a) the third and the fourth output switches (S_{o3}, S_{o4}); (b) the first and second output switches (S_{o1}, S_{o2}); (c) the third output switch (S_{o3}) and the freewheeling switch (S_{fw}).

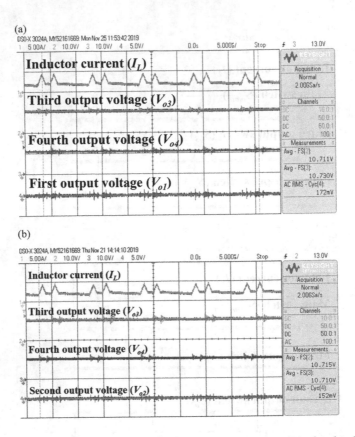

FIGURE 7.19 Measured waveforms of the inductor current (I_L), the third output voltage (V_{o3}), the fourth output voltage (V_{o4}), and (a) the first output voltage (V_{o1}); (b) the second output voltage (V_{o2}).

the second output voltage (V_{o2}) is shown. The experimental results show that the SIFO hybrid converter produces two stable DC output voltages, namely, $V_{o3} = 10.711$ V and $V_{o4} = 10.730$ V, as shown in Figure 7.19a while the first and second (AC) outputs are at zero volts. Hence, it is experimentally verified that standalone DC operation is achievable in the SIFO converter. The total output power of the two DC outputs is measured to be approximately 10.4 W.

7.6.4 Unbalanced Load Condition (Two AC and Two DC Outputs)

An unbalanced load condition is experimentally verified whereby the output power of the AC outputs is different from that of the DC outputs. This can be accomplished by employing *distinct* peak limits of the inductor current (I_L) between the AC and DC outputs while ensuring constant valley limits. In this experiment, the peak limit of I_L for the AC output is *higher* than that for the DC output. Figure 7.20a shows the measured waveforms of I_L, V_{o1}, V_{o2}, and V_{o3} while Figure 7.20b replaces V_{o3} by V_{o4}.

As indicated in Figure 7.20a, the fundamental frequency of the first AC output voltage (V_{o1}) is measured to be around 100 kHz. The measured RMS values of the first and second (AC) output voltages are 10.619 and 10.706 V, respectively. On the other hand, the nominal values of the third and fourth (DC) output voltages are measured to be 9.256 and 9.134 V, respectively. The total output power of the AC outputs is measured to be 10.34 W, whereas that of the DC outputs is 7.69 W. The total output power of all output is around 18 W with a measured efficiency of about 86.3%. As a

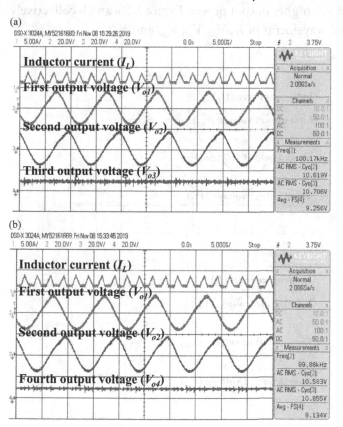

FIGURE 7.20 Measured waveforms of the inductor current (I_L), the first and second output voltages (V_{o1}, V_{o2}), and (a) the third output voltage (V_{o3}); (b) the fourth output voltage (V_{o4}).

step further, let us investigate the possibility of achieving a higher rated output power of the SIFO converter. To safely increase the output power, two requirements must be satisfied: (i) the peak limit of the inductor current must be *lower* than the maximum current rating of the components such as the power MOSFET or the main inductor and (ii) the peak-to-peak inductor current ripple (ΔI_L), which is the difference between the peak and valley current limits, should be adequately small so that the converter can operate safely within PCCM in order to avoid cross-regulation. In this experiment, the average inductor current is increased by raising its peak and valley current limits accordingly while still meeting the above two requirements. Since the average inductor current is the same as the output current for a buck converter, an increase in the average inductor current will lead to a higher output power. Figure 7.21a and b collectively show the measured waveforms of I_L, V_{o1}, V_{o2}, V_{o3}, and V_{o4}.

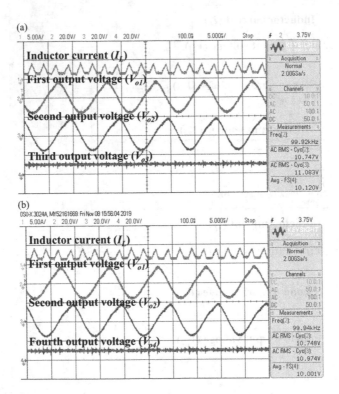

FIGURE 7.21 Measured waveforms of the inductor current (I_L), the first and second output voltages (V_{o1}, V_{o2}), and (a) the third output voltage (V_{o3}); (b) the fourth output voltage (V_{o4}).

The measured RMS values of the first and second output voltages (V_{o1}, V_{o2}) are 10.747 and 11.083 V, respectively [see Figure 7.21a]. The measured DC values of the third and fourth output voltages are 10.120 and 10.001 V, respectively [see Figure 7.21b]. The total output power is increased to 20 W with a measured efficiency of 85.6%.

7.6.5 Practical Hybrid Charger

In this experiment, the SIFO hybrid converter is configured like a practical hybrid charger for simultaneously charging multiple devices. Each AC output is connected to an off-the-shelf transmitting coil that is closely coupled with a loaded receiving coil while each DC output is connected to a pure resistive load. Figure 7.14 shows the whole experimental setup. Basically, the SIFO converter acts as a four-port hybrid converter with two wireless charging pads and two USB charging ports. Given an input power of 25.9 W, the total output power is around 19.67 W. The overall efficiency of the entire system, which includes the two inductive links comprising two pairs of transmitting and receiving coils, is measured to be 76%. Figure 7.22a shows the measured waveforms of the two AC output voltages ($V_{r,\,o1}$, $V_{r,\,o2}$), which are measured at the receiving coils, and the third DC output voltage (V_{o3}). Figure 7.22b replaces the third DC output voltage with the fourth DC output voltage (V_{o4}).

Figure 7.22a and b clearly shows that the AC output voltage at the first (or second) receiving coil is a clean and smooth sine wave. No noticeable distortions or spikes are observed in the measured output voltages. Table 7.3 contains the measured RMS amplitudes of the first harmonic at the fundamental frequency of 100 kHz as well as the second to seventh harmonics. The measured THD is only around 2.454%, which is sufficiently small.

TABLE 7.3   Measured Harmonic Contents of the First Output Voltage Waveform

Frequency (kHz)	Harmonic Order	Measured V_{rms} (V)
100	1	9.79897
200	2	0.237101
300	3	0.034528
400	4	0.016384
500	5	0.008862
600	6	0.005896
700	7	0.007108

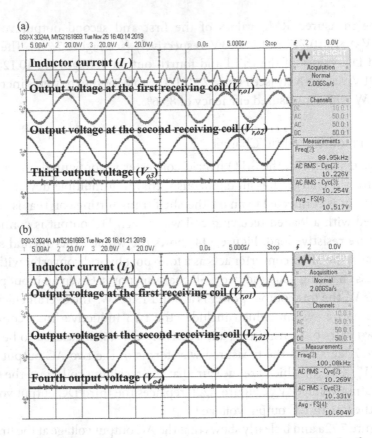

FIGURE 7.22 Measured waveforms of the inductor current (I_L), the first and second output voltages ($V_{r,o1}$, $V_{r,o2}$) measured at the corresponding receiving coils, and (a) the third output voltage (V_{o3}); (b) the fourth output voltage (V_{o4}).

7.7 SUMMARY OF KEY POINTS

1. A single-stage SIMO DC-AC/DC buck hybrid converter is introduced. This hybrid converter employs only a single inductor in the power stage to generate multiple independent AC and DC output voltages. The RMS values of the AC output voltages and the nominal values of the DC output voltages can be individually controlled without any cross-regulation.

2. The conspicuous advantages of this novel SIMO-based hybrid topology are low component count, small form factor, low cost, reduced circuit and controller complexity, high efficiency, and improved

scalability to multiple outputs. This topology is highly versatile and flexible as it allows different combinations of AC and DC outputs. In addition, standalone AC or DC operation is allowed.

3. By disabling all DC outputs, the SIMO DC-AC/DC buck hybrid converter becomes functionally equivalent to a SIMO buck DC-AC inverter. The theoretical analysis shows that this particular buck hybrid converter (or buck inverter) is capable of producing sinusoidal-like AC outputs at each mode of operation.

4. An experimental prototype of a SIFO buck hybrid converter with a maximum rated output power of 20 W is constructed. The experimental results confirm the effectiveness of the SIFO hybrid converter in simultaneously producing two sinusoidal-like AC output voltages and two DC output voltages.

5. In practice, the SIFO buck hybrid converter acts like a multidevice hybrid charger (or power bank) with two wireless charging pads and two USB charging ports for concurrent wireless and wired charging. It supports quick and convenient wireless charging of Qi-certified mobile devices while enabling very rapid charging via USB ports and particularly, providing backward compatibility with older devices without the wireless charging feature.

REFERENCES

1. O. Ray, S. Mishra, A. Joshi, V. Pradeep, and A. Tiwari, "Implementation and control of a bidirectional high-gain transformer-less standalone inverter", *2012 IEEE Energy Conversion Congress and Exposition*, Raleigh, NC, USA, pp. 3233–3240, Sep. 2012.
2. R. Adda, O. Ray, S. K. Mishra, and A. Joshi, "Synchronous-reference-frame-based control of switched boost inverter for standalone DC nanogrid applications", *IEEE Trans. Power Electron.*, vol. 28, no. 3, pp. 1219–1233, Mar. 2013.
3. O. Ray and S. Mishra, "Boost-derived hybrid converter with simultaneous DC and AC outputs", *IEEE Trans. Ind. Appl.*, vol. 50, no. 2, pp. 1082–1093, Mar./Apr. 2014.
4. O. Ray, V. Dharmarajan, S. Mishra, R. Adda, and P. Enjeti, "Analysis and PWM control of three-phase boost-derived hybrid converter", *2012 IEEE Energy Conversion Congress and Exposition*, pp. 402–408, Sep. 2014.
5. A. K. Chauhan, V. R. Vakacharla, M. M. Reza, M. Raghuram, and S. K. Singh, "Modified boost derived hybrid converter: redemption using FCM", *IEEE Trans. Ind. Appl.*, vol. 53, no. 6, pp. 5893–5904, Nov./Dec. 2017.

6. A. K. Chauhan, R. R. Kumar, M. Raghuram, and S. K. Singh, "Extended buck-boost derived hybrid converter", *2017 IEEE Industry Applications Society Annual Meeting*, pp. 1–8, Oct. 2017.

7. V. R. Vakacharla, A. K. Chauhan, M. M. Reza, and S. K. Singh, "Boost derived hybrid converter: problem analysis and solution", 2016 *IEEE International Conference on Power Electronics, Drivers and Energy Systems (PEDES)*, pp. 1–5, Dec. 2016.

8. A. K. Chauhan, M. Raghuram, R. R. Kumar, and S. K. Singh, "Effects and mitigation of nonzero DCM in buck-boost-derived hybrid converter", *IEEE Journal of Emerging and Selected Topics in Power Electronics*, pp. 1470–1482, Sep. 2018.

9. A. Ahmad, V. K. Bussa, R. K. Singh, and R. Mahanty, "Quadratic boost derived hybrid multi-output converter", *IET Power Electron.*, vol. 10, no. 15, pp. 2042–2054, 2017.

10. A. T. L. Lee, W. Jin, S.-C. Tan, and S. Y. R. Hui, "Single-inductor multiple-output (SIMO) buck hybrid converter for simultaneous wireless and wired power transfer", *IEEE Journal of Emerging and Selected Topics in Power Electronics*, (early access), Jun. 2020.

11. D. Ma, W.-H. Ki, and C.-Y. Tsui, "A pseudo-CCM/DCM switching converter with freewheel switching", *IEEE J. Solid-State Circuits*, vol. 38, no. 6, pp. 1007–1014, Jun. 2003.

12. D. Ma and W.-H. Ki, "Fast-transient PCCM switching converter with free-wheeling switching control", *IEEE Trans. Circuits Syst., II: Exp. Briefs*, vol. 54, no. 9, pp. 825–829, Sep. 2007.

13. D. Ma, W.-H. Ki, and C.-Y. Tsui, "Single-inductor multiple-output switching converters in PCCM with freewheel switching," U.S. Patent 7432614 B2, Oct. 07, 2008.

14. F. Zhang and J. Xu, "A novel PCCM boost PFC converter with fast dynamic response", *IEEE Trans. Ind. Electron.*, vol. 58, no. 9, pp. 4207–4216, Sep. 2011.

15. A. T. L. Lee, W. Jin, S.-C. Tan, and S. Y. R. Hui, "Buck-boost single-inductor multiple output (SIMO) high-frequency inverter for medium-power wireless power transfer", *IEEE Trans. Power Electron.*, vol. 34, no. 4, pp. 3457–3473, Apr. 2019.

PROBLEMS

7.1 Draw the circuit diagram of an ideal single-inductor three-output (SITO) boost-type hybrid converter with two AC outputs and one DC output.

7.2 Draw the circuit diagram of an ideal SITO inverting buck-boost hybrid converter with one AC output and two DC outputs.

7.3 Design an SITO boost hybrid converter with two AC outputs and one DC output operating in DCM based on the following design specifications:

Input voltage	$V_{in} = 12\,V$
Main inductor	$L = 2.6\,\mu H$
Load resistor	$R_L = 50\,\Omega$ (assume the same load resistance for each output)

For the AC output:

Resonant capacitor	$C_o = 0.37\,\mu F$
Resonant inductor	$L_o = 6.8\,\mu H$

For the DC output:

Filtering capacitor	$C_f = 6.8\,\mu F$

Suppose this SITO converter employs blocking diodes to eliminate unwanted reverse current flow in order to ensure proper circuit operation.

a. Determine the resonant frequency of the AC output.

b. Determine the switching frequency of the SITO boost hybrid converter.

c. Create a simulation model of an ideal SITO boost hybrid converter based on the above design specification. (Hint: Open loop control may be used to generate the gate driver signals of the power MOSFETs.)

d. Use a simulation software to plot the gate drive signals of the main switch and the three output switches as well as the inductor current and the three output voltages.

7.4

a. For the ideal SITO boost hybrid converter in Problem 7.3, use simulation to find the maximum total output power of the converter without cross-regulation.

b. In a practical SITO converter, the inductor has winding resistance or DCR and the capacitor has equivalent series resistance (ESR). The

DCR of the main inductor and the resonant inductors at their operating frequencies is 20 and 10 mΩ, respectively. The ESR of the resonant capacitors and the filtering capacitor at their operating frequencies is 8 and 12 mΩ, respectively. Besides, the on-resistance of the power MOSFET is 6 mΩ. The forward voltage drop of a Schottky diode is 0.3 V. By incorporating these parasitic resistances and diode's voltage drop into the simulation model of the SITO boost hybrid converter in part (a) and ignoring all other losses, find the maximum total output power that can be achieved by the converter without cross-regulation. Determine the power efficiency of the converter.

7.5 An ideal SIFO buck-type hybrid converter with three AC outputs and one DC output is shown in Figure 7.23.

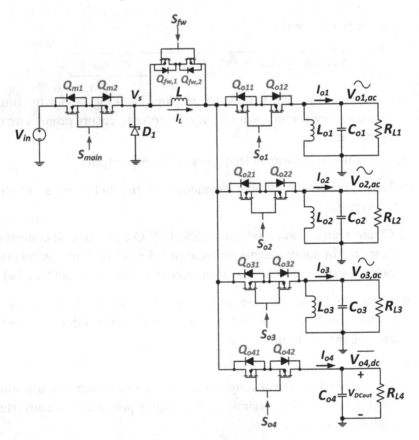

FIGURE 7.23 Ideal SIFO buck-type hybrid converter.

Suppose that the converter operates in PCCM with the following element and parameter values:

Input voltage	$V_{in} = 24\,\text{V}$
Main inductor	$L = 3.3\,\mu\text{H}$
Load resistor per output	$R_{L1} = R_{L2} = R_{L3} = R_{L4} = 22\,\Omega$
Resonant frequency per AC output	$f_o = 100\,\text{kHz}$
Resonant inductor per AC output	$L_{o1} = L_{o2} = L_{o3} = 4.5\,\mu\text{H}$
Resonant capacitor per AC output	$C_{o1} = C_{o2} = C_{o3} = 0.56\,\mu\text{F}$
DC offset of the inductor current	$I_{L,DC} = 4.5\,\text{A}$
Output filtering capacitor of the DC output	$C_{o4} = 33\,\mu\text{F}$

a. Calculate the peak value of the main inductor current for the DC output when the output voltage is 6.76 V, and the on-time duty ratio of the main switch (S_{main}) is 7%.

b. Suppose the on-time duty ratio of the fourth output switch (S_{o4}) is 25%. Determine the average value of the DC output current (I_{o4}).

c. Create a simulation model of the ideal SIFO buck hybrid converter based on the above design specification. Use simulation to plot the waveforms of the main inductor current, the three AC output voltages, and the DC output voltage.

d. Use the simulation model in part (c) to show that the SIFO buck hybrid converter is capable of operating in standalone AC mode (i.e., only the first three AC outputs are enabled). Plot the waveforms of the main inductor current, the three AC output voltages, and the DC output voltage.

7.6 The SIFO buck hybrid converter with two AC outputs and two DC outputs operating in PCCM contains a total of 12 power MOSFETs as shown in Figure 7.4.

a. Optimize this hybrid converter by using fewer MOSFETs. Sketch the resulting circuit topology. State the condition for proper circuit operation of the optimized converter.

b. The design optimizations in part (a) can also be extended to a general SIMO buck hybrid converter operating in PCCM. Derive a general expression for the total number of power MOSFETs of the optimized SIMO buck hybrid converter in terms of the total number of AC outputs (N_{ac}) and the total number of DC outputs (N_{dc}).

c. Based on the expression derived in part (b), determine the total number of power MOSFETs required in an SIFO buck hybrid converter with two AC outputs and two DC outputs.

7.7 Consider a nonideal SIFO buck hybrid converter with two AC outputs and two DC outputs whose design specifications are given below.

Input voltage	$V_{in} = 24\,V$
Switching frequency	$f_{sw} = 400\,kHz$
Output resonant frequency	$f_o = 100\,kHz$
Main inductor	$L = 3.3\,\mu H$
DCR of the main inductor	$r_L = 50\,m\Omega$ (measured at 400 kHz)
Voltage drop of the freewheeling diode	$V_d = 0.3\,V$
On-resistance of the power MOSFET	$R_{on} = 5\,m\Omega$
Resonant capacitor	$C_{o1} = C_{o2} = 0.56\,\mu F$
ESR of the resonant capacitor	$r_{Co1} = r_{Co2} = 8\,m\Omega$ (at 100 kHz)
Resonant inductor	$L_{o1} = L_{o2} = 4.5\,\mu H$
DCR of the resonant capacitor	$r_{Lo1} = r_{Lo2} = 10\,m\Omega$ (at 100 kHz)
Output filtering capacitor of the DC output	$C_{o3} = C_{o4} = 22\,\mu F$
ESR of the output filtering capacitor	$r_{Co3} = r_{Co4} = 12\,m\Omega$ (at 100 kHz)
Load resistor per output	$R_{L1} = R_{L2} = R_{L3} = R_{L4} = 22\,\Omega$
Mode of operation	PCCM
DC offset of the inductor current	$I_{L,\,DC} = 2.5\,A$

All other losses can be ignored.

a. Construct the simulation model of this nonideal SIFO buck hybrid converter using back-to-back connected MOSFETs as illustrated in Figure 7.4. Use the quasi-hysteretic feedforward control scheme to regulate the average inductor current for each output.

b. Suppose the main inductor current has a uniform peak limit of 4.0 A across all outputs while its valley limit remains constant at 2.5 A. Use simulation to plot the main inductor current, the two AC output voltages, and the two DC output voltages.

c. Based on the simulation results in part (b), obtain the RMS values of the two AC output voltages and the average values of the two DC output voltages. Explain why the RMS value of the AC output voltage differs from the average value of the DC output voltage even though the same peak (or valley) limits are used across all outputs.

d. Determine the power efficiency of the SIFO converter from the simulation results in part (b). In this case, what is the major contributor to the overall power loss? Explain why.

Index

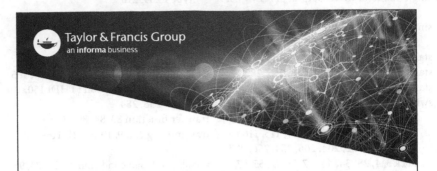

Printed in the United States
by Baker & Taylor Publisher Services